A-Z
PHYSICS

A-Z
PHYSICS

Prof. G.K. Bose

CENTRUM PRESS
NEW DELHI-110002 (INDIA)

CENTRUM PRESS
H.O.: 4360/4, Ansari Road, Daryaganj,
New Delhi-110 002 (India)
Ph.: 23278000, 23261597

B.O.: No. 1015, Ist Main Road, BSK IIIrd Stage
IIIrd Phase, IIIrd Block,
Bangalore - 560 085 (India)
Tel.: 080-41723429
Visit us at: www.centrumpress.com

A-Z Physics

First Edition, 2009

ISBN 978-93-80106-32-8

PRINTED IN INDIA

Printed at Salasar Imaging Systems, Delhi-110035 (India)

Contents

Preface

Physics, more than any other science, can be demonstrated principle after principle by direct and simple experiments. The implements of the present attack upon the basic principles of our physical knowledge are demonstrable. The topics that should be included in the physics curriculum and the depth of each topic have to be selected very judiciously. For example, Newton's Laws are fundamental, and they are discussed in detail both theoretically and in numerous applications. On the other hand, some topics of modern physics had to be considered secondary, and they are interwoven at a qualitative level within other chapters. It should be noted, that "A-Z Physics" is a physics textbook, and it is not intended to be a replacement for the professional architecture books.

Although this book has been designed for graduate or undergraduate students, of physics. The physics principles, concepts, formulas and ideas brought here are the same as those presented in other general purpose, introductory physics textbooks. The main difference is in the examples used to introduce and to explain the physics. Those examples are related to buildings and structures, and to their use by humans. Readers from other disciplines too may find the book interesting and effective in their study of physics.

Author

Chapter 1

Physics - An Introduction

Physics is often described as the study of matter and energy. It is concerned with how matter and energy relate to each other, and how they affect each other over time and through space. Physicists ask the fundamental questions how did the universe begin? how and of what is it made? how does it change? what rules govern its behaviour?

Physicists may be roughly divided into two camps: experimental physicists and theoretical physicists. Experimental physicists design and run careful investigations on a broad range of phenomena in nature, often under conditions which are atypical of our everyday lives. They may, for example, investigate what happens to the electrical properties of materials at temperatures very near absolute zero (-460 degrees Fahrenheit) or measure the characteristics of energy emitted by very hot gases.

Theoretical physicists propose and develop models and theories to explain mathematically the results of experimental observations. Experiment and theory therefore have a broad overlap. Accordingly, an experimental physicist remains keenly aware of the current theoretical work in his or her field, while the theoretical physicist must know the experimenter's results and the context in which the results need be interpreted.

It is also useful to distinguish classical physics and modern physics. Classical physics has its origins approximately four hundred years ago in the studies of Galileo and Newton on mechanics, and similarly, in the work of Ampere, Faraday, Maxwell and Oersted one hundred fifty years ago in the fields of electricity and magnetism. This physics handles objects which are neither too large nor too small, which move at relatively slow speeds (at least compared to the speed of light: 186,000 miles per second!). The emergence of modern physics at the beginning of the twentieth century was marked by three achievements.

The first, in 1905, was Einstein's brilliant model of light as a stream of particles (photons). The second, which followed a few months later, was his revolutionary theory of relativity which described objects moving at speeds close to the speed of light. The third breakthrough came in 1910 with Rutherford's discovery of the nucleus of the atom.

Rutherford's work was followed by Bohr's model of the atom, which in turn stimulated the work of de Broglie, Heisenberg, Schroedinger, Born, Pauli, Dirac and others on the quantum theory. The avalanche of exciting discoveries in modern physics continues today.

Given these distinctions within the field of physics experimental and theoretical, classical and modern it is useful to further subdivide physics into various disciplines, including astrophysics, atomic and molecular physics, biophysics, solid state physics, optical and laser physics, fluid and plasma physics, nuclear physics, and particle physics.

NEWTON'S THREE LAWS OF MOTION

The most fundamental concepts in physics were derived by the great English scholar Isaac Newton in the 17th century, in the form of what are now known as Newton's three laws of motion. These three simple principles describe the motion of objects and underlie all modern physics. This chapter provides an introduction to these laws and demonstrates how they relate to the concepts of "work" and "energy".

GALILEAN RELATIVITY / NEWTON'S THREE LAWS OF MOTION

One of the first questions addressed in the development of physics was: Why does some object keep moving even after the force applied to it goes away? If a cannonball is blasted out of the mouth of a cannon, it flies through the air in a smooth curve until it strikes something or is pulled down to the ground by gravity. The answer to this question was finally unraveled by the Italian scientist Galileo Galilei. Galileo saw through the question by seeing that the concept of motion was strictly relative: it is arbitrary to say that A is in motion relative to B, or the reverse.

This sounds like a deep and exotic concept, but the idea is very simple. Suppose Dexter's on a ship, inside a closed cabin, sailing on a glassy smooth sea. If he doesn't look outside, there is no way for him to know if he is moving or not. The physics of everything in the room would be exactly the same while the ship was in uniform motion as it would if the ship was sitting at the dock. This can be broadly referred to as the "relativistic equivalence principle".

The idea that there was an absolute state of rest was a natural mistake, since we live on a large ball of rock called the Earth that is a lot bigger than we are, bigger than anything that we can set into motion. This means that the Earth seems very convincingly at rest, and anything that moves relative to it seems very convincingly in motion.

It is much easier to see the relativity of motion by imagining Dexter floating in space inside a glass box. He sees his sister Deedee flying past him in a glass box of her own. As far as Dexter is concerned, it is Deedee who is in motion. However, from the point of view of Deedee in her glass box, it is she who is motionless and it is Dexter who is moving, in the opposite direction. Dexter and Deedee reside in what are referred to as different "inertial frames of reference".

Modern generations have realized that the Earth is itself in motion around the Sun, and the Sun is in motion around the centre of our Galaxy.

Such cosmic concepts of motion are still hard to grasp intuitively, and so in practice we invariably think of something sitting motionless on the ground as being "at rest" while a car driving over the ground is "in motion" The Earth defines a "local" frame of reference for our motions, and it would be silly to think otherwise. It is useful for someone interested in physics to keep in the back of his or her mind that our Earth-centered viewpoint is a parochial point of view, and in other inertial frames of reference it is the Earth that is in motion.

- Galileo's consideration of the meaning of rest and the

meaning of motion sounds a little like doubletalk, but it did have practical implications. The main implication was that motion itself was less important than changes in motion, or accelerations. A force causes a change in motion of an object, and once set into motion that object will remain on that path of motion, until another force speeds it up, slows it down, changes its direction, or causes any other change in its motion.

Galileo laid the groundwork, but never managed to link all the ideas together; that task was performed after him by the English genius Sir Isaac Newton in his three laws of motion.

- *First Law*: Objects at rest remain at rest, and objects in motion remain in motion in a straight line at constant velocity. A force must be applied to change the state of motion of an object.
- *Second Law*: The acceleration of an object is directly proportional to the force acting on it, and inversely proportional to its mass.
- *Third Law*: If two objects interact, say in a collision, the force exerted by object 1 on object 2 is matched by a force of exerted by object 2 on object 1 of the same size, but in the opposite direction.

These three simple, elegant laws are adequate to describe the motion of all objects in the Universe.

THE FIRST AND SECOND LAW

- The First Law observes that objects in motion tend to remain in motion in a straight line. The magnitude of this tendency is given by a property called "momentum", which is unsurprisingly related to the bulk or "mass" of an object, and the velocity with which it is moving. In mathematical terms, the First Law is expressed as:

$$\text{Momentum} = \text{mass} \times \text{velocity}.$$

In the metric system, mass is by default measured in kilograms. Momentum, like velocity, is a vector, and has a magnitude and a direction. The Second Law defines force and mass. In mathematical terms, the Second Law is expressed as:

$$\text{Force} = \text{mass} \times \text{acceleration}$$

What is a force? It's what accelerates a mass. What's a mass? It's what is accelerated by a force. Gravity, for example, is a force that accelerates a mass such as a brick. Force and mass are fundamental concepts, basic definitions that have to be taken as givens. In the metric system, force is measured in "newtons".

Since acceleration is, in broad terms, the change or "delta" of velocity over an increment or "delta" of time, and momentum is mass times velocity, with the mass being constant:

Force = mass × acceleration
= mass×delta velocity / delta time

and:

Force = delta momentum / delta time

This is another way of stating the Second Law, or in other words force is equivalent to the rate of change of momentum over time: the greater the force, the greater the rate of change of momentum.

- Incidentally, mass and weight are not the same thing. A mass remains constant no matter where it is in the Universe. In contrast, weight is just a measure of the force of gravity on a mass. If Dexter goes to the Moon, he would only weigh a sixth of what he weighs on Earth, but his mass would be the same.

Modern physics recognizes that the "matter" in the Universe that constitutes its mass is made up of vast numbers of very small particles, known as "atoms", that as a general (but by no means strict) rule cannot be subdivided further. There are a roughly a hundred different types of atoms that have different properties. Atoms can combine to form a wide range of "molecules" that have different properties from the atoms that make them up.

Atoms and molecules can be organized as solids, liquids, or gases. While we live on Earth completely surrounded by matter, it is actually relatively scarce in the Universe. Once we travel away from the Earth into space, the amount of matter in a volume of space becomes effectively negligible. Space empty of appreciable amounts of gaseous matter is known as a "vacuum"; most of the Universe is a vacuum.

- Some examples can help demonstrate what the First and Second Laws mean in practice:

- To go back to the idea that started this discussion, suppose Dexter fires a cannonball out of a cannon. The cannonball is put into motion by the force of the explosion. If he's firing the cannon out of a spacecraft in empty space, far from any planet or star, the cannonball will go on forever at the same speed, in a straight line. The cannonball's property of remaining at rest until put into motion and then remaining in motion is called "inertia".

If Dexter fires the cannon on the surface of the Earth, the Earth's force of gravity is acting on it, and so it drops as it flies forward until it hits the ground. The force of gravity is a vector that pulls straight down while (in this example) the velocity of the cannonball is a vector pointed flat over the ground. The gravitational force doesn't change the horizontal velocity at all, it just adds a vertical velocity to it as well.

If Dexter fires a cannonball on a flat trajectory out of a cannon and drops a cannonball from muzzle height at the same time, both cannonballs hit the ground at the same time. It falls down just as fast if it is moving horizontally as it does if it dropped straight down. Ignoring drag, it will be moving horizontally just as fast as it did when it left the muzzle.

Although this is getting a bit ahead of the discussion, it should be noted that this scenario assumes the range of the cannon is fairly short. If the range is long, however, the cannonball flies over the horizon.

The Earth curves out from underneath the cannonball as it flies, making the distance it has to drop to hit ground longer. In fact, if Dexter fired the cannonball fast enough, the Earth would curve out underneath it so quickly that the cannonball would never hit ground, and the cannonball would orbit completely around the Earth. To rephrase the British humorist Douglas Adams: The secret of orbiting is to fall to the ground and miss.

This scenario assumes that there is no atmosphere. The atmosphere causes drag against the flight of the cannonball, creating a force that increases with its velocity and operates against the direction of the cannonball's flight, slowing it down and causing it to hit the ground in time. If there were no atmosphere, the cannonball could orbit the Earth all but indefinitely.

- There is a popular belief that a force has to be applied to keep an object in motion. This is, to an extent, just common sense: if Dexter is flying a jet fighter and the engines flame out, it won't fly any more it's going down. However, common sense misses the important fact that the force of drag is continuously opposing the fighter's motion, and a continuous force is required to overcome that drag.
- Now suppose Dexter is in a spacecraft flying through empty space. If he fires his rocket engine, the spacecraft will speed up; when he turns off the engine, the spacecraft will keep moving at the same velocity indefinitely, in a straight line.
- Imagine Dexter is driving a pickup truck with a bowling ball rolling around loose on the truck bed. If Dexter steps on the gas, the bowling ball rolls back on the truck bed and hits the tailgate. However, the bowling ball is not moving so much as the truck is, with the bowling ball remaining in its original position. Once it hits the tailgate, the truck is providing a force that accelerates the bowling ball to the same speed as the truck.
- Similarly, if Dexter turns left, the bowling ball slams into the right side of the truck bed; if he turns right, it slams into the left side of the truck bed; and if he stops, it slams into the front of the truck bed, since the bowling ball keeps on moving until a force makes it stop. Dexter then gets out of the truck to put the bowling ball in the cab before it dents the truck into a wreck.
- Another angle on this is to consider what happens if Dexter is driving along in his truck with a cup of water in his hand. If he's driving along steadily on a smooth road, the liquid in the cup will be just as level as if he was standing still. It's only if he speeds up, turns, or slows down that the liquid in the cup will slosh backward, to the side, or to the front respectively.

Try putting a small sealed jar of water on the dashboard of your car to see this in action. Consider what happens if Dexter drops an object while he's running. There is another popular belief that the object will fall straight down to the ground at the point where he dropped it, but the First Law shows that's not the case. Ignoring drag, which is a reasonable assumption if the item is compact like a rock, it moves alongside him until it hits the ground.

This is a fact of life for combat aircraft dropping bombs at low altitude. The bombs, which are generally streamlined to reduce drag and allow the aircraft carrying them under its wings to fly faster, will fall forward and remain under the aircraft, which will be caught in the blast and destroyed. This is why bombs configured to be dropped at low altitude pop out paddles, small parachutes, or other types of "retarders" to slow them down through increased drag, so the aircraft can escape the blast.

WORK AND ENERGY

- Closely related to the concept of force are the dual concepts of "work" and "energy". Work is defined as the sum of forces applied to a mass over a distance. Energy is the capacity to do work.

The foundations of these concepts were laid in Newton's work, as well as in the work of the German polymath Gottfried Wilhelm Leibniz, a rival of Newton. The notions were more clearly defined by the experiments of the Dutch physicist Willem s'Gravesande, who performed meticulous studies of metal balls on strings bouncing against each other, spherical masses dropped on beds of clay from different heights to see how big of dent they made, and so on.

In the 19th century, the notions of work and energy were investigated in detail by the British physicist James Prescott Joule and the German physicist Julius Robert von Mayer, leading to the precise formulation of modern concepts of work and energy by another German physicist, Hermann Ludwig Ferdinand von Helmholtz. For a simple example of work and energy in action, imagine that Dexter lifts a mass against the Earth's gravity. In his immediate surroundings, the force of the Earth's gravity is for all practical purposes constant. Actually, gravity grows weaker the farther an object is away from the Earth, but for such short distances the force of gravity varies too little to notice.

Since gravity is exerting a force on the mass, he has to exert a force on the mass to lift it against gravity. If he lifts the mass to, say, three meters, the total work performed on the mass is the force times three meters. At least in terms of pure physics, how long it takes him to lift the mass doesn't matter, nor do any side trips taken on the way up; that all amounts to wasted motion, what is important is that it is lifted three meters.

If the mass is raised to that height, it then has a capability to do work itself by falling back down again. This is the "potential energy" of the mass, and is the same as the amount of work required to raise it that height. The potential energy of a mass being raised near the surface of the Earth is given by:

Potential Energy = 9.81 × mass × height

Energy and work are measured in "joules", named after James Prescott Joule. If a mass of ten kilograms is raised three meters, its total potential energy is:

Potential Energy = 9.81 × 10 × 3 = 294 joules

The rate at which work is done or energy is expended is defined as "power", which is measured in "watts", or joules per second. If Dexter lifts the weight in five seconds, the power used to do it is:

Power = 294 / 5 = 58.9 watts

Of course, inversely, multiplying the power used times the time it is applied gives the total energy.

Energy = Power × time = 5×58.9 = 294 joules

This definition leads to an alternate unit for energy named "kilowatt-hours", often used in electrical power distribution. A kilowatt-hour is a thousand watts applied for an hour, or 3,600 seconds, and so is equivalent to 3,600,000 joules.

- Moving objects also store energy due to their motion. A flying brick, for example, has the capability of doing destructive work. This form of energy is known as "kinetic energy".

The kinetic energy is the sum of all the force provided over a distance needed to get the brick to a specific velocity. Consider a simple example, where a mass is accelerated from rest by a constant acceleration for a specific time interval. As shown earlier, for a constant acceleration, the average velocity is half the final velocity V, so the total distance D traveled in the given time T is:

$$D = (1/2) \times V \times T$$

Now since:

$$\text{Force} = \text{mass}\times\text{acceleration} = M \times A$$

and since:

$$\text{Energy} = \text{force}\times\text{distance}$$

then for this case:

$$\begin{aligned}\text{Energy} &= M \times A \times D \\ &= M \times A \times (1/2) \times V \times T\end{aligned}$$

Since in this case the acceleration is constant, then the acceleration is the final velocity V divided by the time interval T, and so:

$$\begin{aligned}\text{Energy} &= M \times (V/T) \times (1/2) \times V \times T \\ &= (1/2) \times M \times V^2\end{aligned}$$

This gives the definition of kinetic energy as:

$$\text{Kinetic energy} = (1/2) \times \text{mass} \times \text{velocity}^2$$

Kinetic energy and momentum have a simple relationship:

$$\text{Kinetic energy} = (\text{momentum}^2) / (2 \times \text{mass})$$

since:

$$\begin{aligned}\text{Kinetic energy} &= ((M \times V)^2) / (2 \times M) \\ &= (M^2|) \times (V^2) / (2 \times M) \\ &= (M \times V^2) / 2\end{aligned}$$

Kinetic energy, and in fact any sort of energy, is a scalar, not a vector, quantity: it has a magnitude, but no direction. While this derivation is only for a constant acceleration, the kinetic energy only depends on the mass and velocity. How the mass got to that velocity doesn't matter, and so the definition is true in the general case as well.

By the way, since momentum is given as mass times velocity, this expression can also be redefined as:

$$\text{Kinetic Energy} = (1/2) \times \text{momentum} \times \text{velocity}$$

- Notice that kinetic energy is proportional to the mass and to the *square* of the velocity. This means that even a small mass moving very fast has a lot of kinetic energy. Suppose Wile E. Coyote is hit by a 1 tonne (1,000 kilogram) car moving at 100 kilometers per hour (or 100 × 1,000 / 3,600 = 27.8 meters per second). Then the total kinetic energy is:

$$\text{Kinetic energy} = (1/2) \times 1{,}000 \times 27.8^2 = 385{,}800 \text{ joules}$$

Let's consider how fast a 10 gram mass would have to be moving to have this same kinetic energy:

$$\text{Velocity} = \sqrt{(2 \times \text{Kinetic energy / mass})}$$

$$= \sqrt{(2 \times 385{,}800 / 0.01)} = 8{,}780 \text{ meters}$$

per second

= 31,600 KPH

- Energy is a concept now taken so much for granted that it is startling to realise that Isaac Newton didn't have a clear concept of the idea.
- It was certainly apparent to Newton that a stone sitting on top of a high wall had a capability to do damage to someone standing below if it were knocked off, and of course everyone understood that a flying cannonball could be dangerous indeed, that was the *point* of cannonballs.

Leibniz was the first to articulate the idea, coming up with the terms "vis visa" and "vis mortua", which were direct ancestors of the modern definitions of kinetic and potential energy respectively, but the word "energy" itself wasn't introduced until 1807, when it was suggested by the brilliant British polymath Thomas Young. It would take several more decades to refine the notion into its modern form. Energy is a highly abstract concept: motion and mass are easily observed, but it's not like anyone can actually *see* energy. From that point of view, it is also surprising to realise that in modern times, energy is something we can actually measure to a high degree of precision.

CONSERVATION OF ENERGY

- Energy is said to be a "conserved" quantity. Wile E. Coyote exerts work to haul an anvil to the top of a cliff, giving the anvil potential energy, and this potential energy is then converted back into kinetic energy when the anvil falls back down to the bottom of the cliff, with friction effects causing losses along the way.

This basic principle is, as far as anybody has ever seen, absolutely unbreakable: energy is conserved. It cannot be created or destroyed, just converted into different forms. The energy "budget" before a physical event occurs remains exactly the same after the event occurs; only the forms are changed. Helmholtz was the first to articulate this concept in its modern form, though he referred to kinetic energy as "living" energy and potential energy as "tensional" energy.

The law of conservation of energy is why most perpetual motion machines won't work. Typical "classic" perpetual motion machines operate in a cycle, with an output feeding back to an input. The Dutch artist Maurits Escher was fond of making paper designs along such lines, such as water flowing in a loop and driving a waterwheel.

Escher's work was deliberate and obvious fantasy, playing "crazy house" games with perspective; everybody knows that water doesn't flow back uphill. The perpetual motion machines created by crackpots are usually just as absurd, with most of the effort in the exercise focused on trying to conceal the absurdity. No matter how elaborate the disguise, however, the idea that a machine can do useful work indefinitely without input of energy is a dead giveaway. Perpetual motion machines are almost always paper designs, since when they're actually built, they usually don't move at all.

- Back in the real Universe, one simple example of the law of conservation of energy is to catapult a mass into the sky and determine how high it goes.

  Suppose Dexter catapults a 1 kilogram mass upward at an initial velocity of 30 meters per second. Then the initial kinetic energy is:

$$\begin{aligned}\text{Kinetic energy} &= (1/2) \times \text{mass} \times \text{velocity}^2 \\ &= (1/2) \times 1 \times 30^2 \\ &= (1/2) \times 900 \\ &= 450 \text{ joules}\end{aligned}$$

As the mass rises, gravity will slow its rise, converting the kinetic energy into potential energy. Ignoring drag, the mass will stop at the height at which the potential energy equals the initial kinetic energy:

$$\begin{aligned}\text{Potential Energy} &= 9.81 \times \text{mass} \times \text{height} \\ 450 &= 9.81 \times 1 \times \text{height} \\ \text{Height} &= 450 / 9.81 = 45.9 \text{ meters}\end{aligned}$$

Having reached this height, the mass will then begin to fall back down again. When it reaches the altitude it was launched from, it will once again be moving at 30 meters per second. mIncidentally, an alert reader will realise that the time it takes for the mass to fall back down is given by exactly the same formula used to describe how long it takes an anvil to fall on the Coyote it's falling from a stopped position either way:

$$\text{Time} = \sqrt{(2 \times \text{displacement} / 9.81)}$$

$$= \sqrt{(2 \times 45.9 / 9.81)}$$

$$= 3.06 \text{ seconds}$$

Since it takes the mass the same amount of time to fly up as it does to fall back down the trip up is exactly the same as the trip down, run in reverse the total flight time is 2×3.06 = 6.12 seconds.

- Very much the same principles apply to a roller coaster. A classic roller coaster is essentially unpowered, with the cars hauled to the top of a "lift hill" by a chain drive built into the hill, and then allowed to coast freely down the other side over a series of smaller hills, and often (with modern steel-rail roller coasters) through loops and corkscrews.

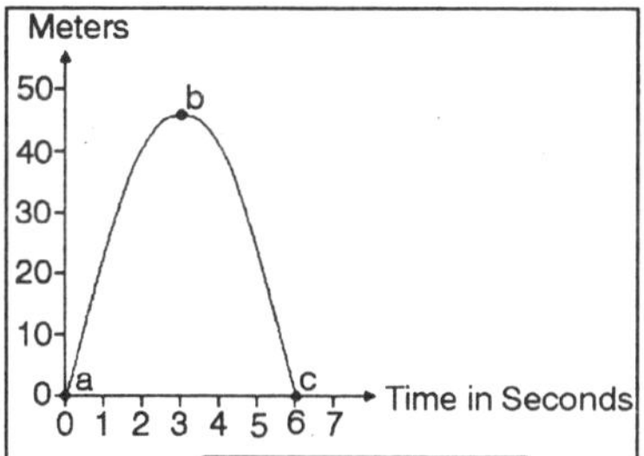

Conservation of energy

As the cars descend down a hill they pick up speed and kinetic energy; lose it again as they rise to the top of the next hill, acquiring potential energy; pick up speed and kinetic energy once more as they drop down the far side; and so on, with the energy gradually robbed away by friction until the cars finally come to a stop and the passengers stagger out. The initial amount of energy applied to the cars is directly proportional to the height of the lift hill, and in fact the height of the lift hill is the most important single specification for a given roller coaster to an enthusiast.

Inverted Roller Coaster (GVG/PD)

Another interesting simple example of the law of conservation of energy is the kind of spring-loaded toy gun that shoots little plastic spheres. If Dexter pulls back on the trigger of the gun, then up to the limit of the complete compression of the spring, the force he has to exert on the trigger is proportional to the stiffness of the spring and increases directly with the compression of the spring. In mathematical terms:

$$\text{Force} = \text{Stiffness} \times \text{Compression}$$

In general terms, the stiffer the spring, the harder it is to pull the trigger, and the longer the pull on the trigger, the harder it is to pull further. This relationship is known as "Hooke's law", after the English scientist Robert Hooke, a contemporary of Newton.

Dexter does work pulling back on the trigger, with the work being stored up in the spring as potential energy.

The work is equal to the force and the distance over which the force is applied. Since the force increases in a nice straight line over the distance of compression, the average force over the entire compression is just half the final compression, so the potential energy is:

$$\text{Potential Energy} = (1/2) \times \text{stiffness} \times \text{compression}) \times \text{compression}$$
$$= (1/2) \times \text{stiffness} \times \text{compression}^2$$

When Dexter reaches the end of the trigger pull, the spring mechanism is automatically released and the spring returns to its normal length, exerting force on the sphere to accelerate it out of the muzzle, and converting the potential energy into motion, or equivalently kinetic energy. By the way, Hooke's law also applies to stretching a spring, at least up to the limit where it "deforms", or stretches out of shape and won't return to its original form when released.

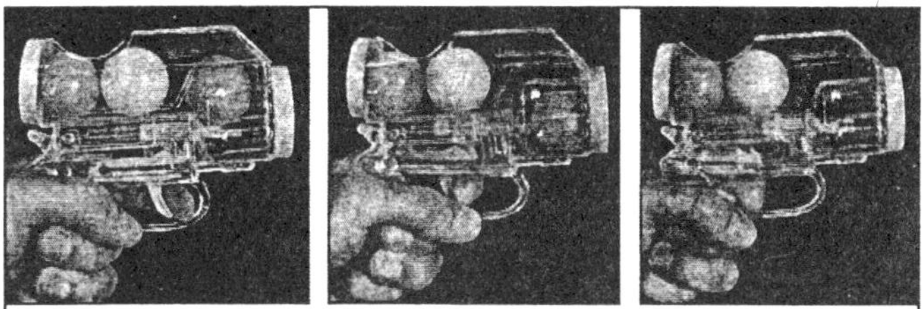

Spring-Operated GUN (GVGPD)

- As an interesting footnote, Hooke and Newton were bitter rivals. Newton was a brilliant but not always very pleasant person, with a tendency to be resentful of and ungracious to competitors. In a 1676 letter, Newton famously told Hooke: "If I have seen further [than others], it is by standing on the shoulders of giants." This has traditionally been cited as evidence of Newton's modesty, but in fact Hooke was an unusually short person, and in modern times the consensus is that it was more likely that Newton was really being snide, saying that he owed nothing in his own work to a nasty little runt like Hooke.

THE THIRD LAW / MOMENTUM IN ACTION

- The Third Law describes, among other things, action and reaction. If Dexter fires a gun, the bullet flies out the barrel and the gun recoils or kicks back in the opposite direction. Both the bullet and the gun have a certain value of momentum, and in fact it's the same value, though the directions are opposite:

Mass of bullet ×velocity of bullet = mass of gun ×velocity of gun

This means that the velocity of the gun is slower than the velocity of the bullet by the same factor that the mass of the gun is greater than the mass of the bullet. If the gun has 100 times the mass of the bullet, it only moves 1/100th as fast in the opposite direction. In more general terms:

$$\text{mass}_1 \times \text{velocity}_1 = \text{mass}_2 \times \text{velocity}_2$$

A more ordinary way to think about the third law is to consider what happens if Dexter and Deedee are standing on slick ice next to each other, and then push each other away. Dexter doesn't just stand still while Deedee slides away they both slide away from the point they were at. If they're both about the same mass, they slide away in opposite directions at the same speed. If Dexter weighs only half as much as Deedee, he slides away at twice the speed of Deedee.

- The essential point here is that the momentum of a system is a conserved quantity. This is seen in collisions between moving masses, most conveniently visualized as spheres. Such collisions may be "elastic" or "inelastic". In simple discussions of momentum, it is easiest to specify it in terms of one-dimensional movements along a straight line in one direction or the other. We can consider one-dimensional examples of elastic and inelastic collisions.

In an elastic collision, the kinetic energy of the masses is preserved. This is the case if the masses are very hard and stiff. In a head-on elastic collision between two spheres of the same size and same velocity in different directions, the two spheres bounce away from each other at their original velocity but with reversed directions. Total momentum is conserved.

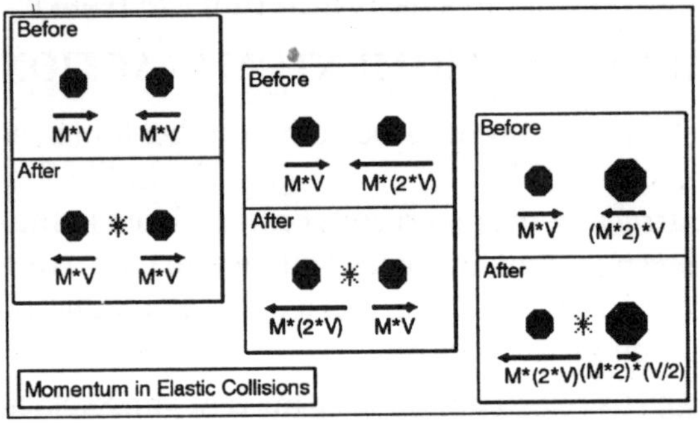

If the masses or velocities are not the same in an elastic collision, the results are little more interesting. Suppose one mass is moving twice as fast as a second mass of the same size, and so has twice the momentum. If there is an elastic head-on collision between the two, they bounce away from each other, but the second mass is now moving twice as fast as the first. This occurs because momentum is a vector quantity, and has to be conserved in direction as well as size.

Now suppose there are two masses that are moving at the same speed, but the first mass is twice as big as the second and so has twice as much momentum. If there is an elastic head-on collision between the two, they bounce away from each other, but the conservation of momentum says that the big mass will bounce away at only half its initial speed, while the small mass bounces away at twice its initial speed.

- In an inelastic collision, the kinetic energy of the masses is converted into heat. This happens if the masses are soft and malleable. In a head-on inelastic collision between two spheres of the same mass and same velocity in different directions, the two spheres slap into each other and come to a dead stop. Momentum is conserved once more.

Of course, if the momentum of one mass in such an inelastic collision is greater than the second, its momentum will overwhelm that of the second, and the combined mass that results from the collision will move in the direction of the first. Suppose the first mass is twice the size of the second, but both are moving at the same velocity.

Then the total momentum of the resulting particle has the same size but opposite direction as the momentum of the second particle. Since the mass of the resulting particle is three times the size of the second particle, the velocity is cut to one-third of the original.

Similarly, if the velocity of the first particle is twice that of the second, the resulting momentum is still the same. However, the resulting particle is twice the mass of the second, so the velocity is only cut in half.

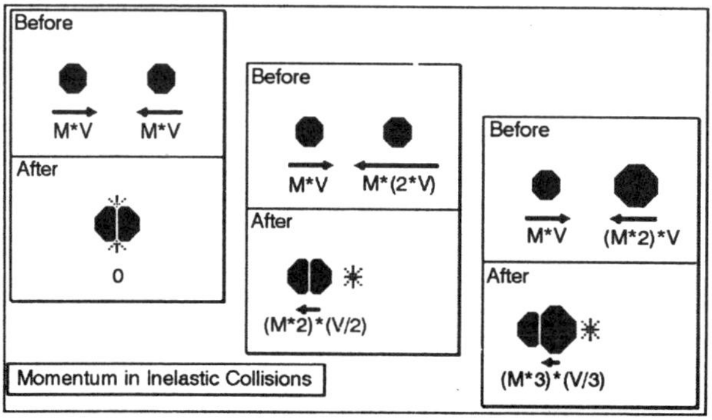

Momentum in Inelastic Collisions

- It is useful to realise that perfectly elastic collisions don't actually occur in the real world: all collisions will have some inelastic component, however small, though in a simple analysis that component may be small enough to be ignored without difficulty. Physics, particularly simple and basic physics, deals with idealized "models" that will not be quite what happens in the imperfect real world, though they may be close enough for practical purposes.

Such simplifications are not done out of any intent to deceive, only out of a desire to make a problem less complicated so it can be easily analyzed. It took pioneering physicists great leaps of intuition to see through the clutter of the real world and extract useful basic principles of physics.

It is also useful to realise that sometimes the simplified principles can be wrong or at least misapplied, though all the physics in this document is very simple and thoroughly validated through observation and experiment which is always the ultimate test of any theory in the sciences.

- By the way, the law of reaction is very well illustrated by rocket propulsion. There is a strange popular misconception that a rocket operates by "pushing against the air", but that is not the case. What a rocket does is throw hot gas out the exhaust at high velocity, and by conservation of momentum the rocket moves in the other direction. Essentially, a rocket propels itself using recoil. The air just gets in the way and causes drag.

The more mass in the exhaust and the faster it is sent out, the more thrust it creates and the faster it propels the rocket. In more formal terms, if Me is the mass of the exhaust per second, Ve is the velocity of the exhaust, Mr is the mass of the rocket, and Vr is the velocity of the rocket, then:

$$Me \times Ve = Mr \times Vr$$

Calculating the velocity of a rocket can be tricky since its mass drops as it burns and exhausts fuel, but that is a matter far beyond the scope of an elementary physics document.

Of course, as mentioned, the Third Law is demonstrated very well by the kick of a rifle or cannon. The recoil increases with the size of the shell being fired -- more mass, increasing momentum and the size of the charge firing it more velocity, also increasing momentum.

Really powerful rifles can generate quite a kick. Most modern assault weapons use a relatively small cartridge and bullet, reducing recoil, and often have a spring-loaded buffer system in the stock to spread out the time over which the momentum is transferred back to the shooter, making the kick much less abrupt and softer.

Such weapons are easier to fire without extensive training, and also much more controllable in full automatic ("machine gun") fire. Similarly, modern artillery pieces often have spring or hydraulic suspensions to soften the recoil, along the lines of the suspensions used in vehicles to soften the effect of bumps in the roadway.

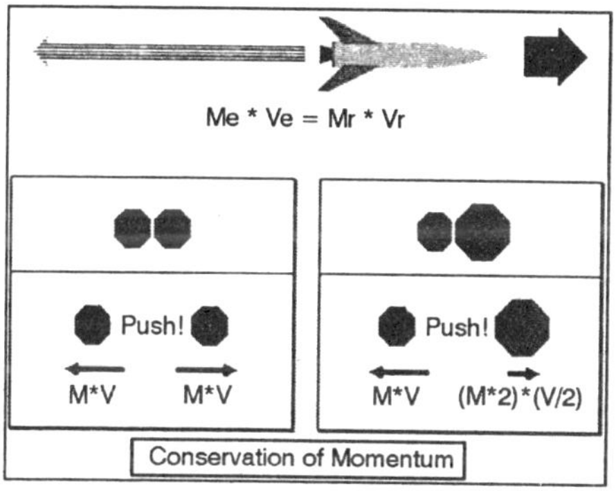

Trying to fit a heavy cannon with suspension gear in an aircraft is troublesome. That's why really big guns have rarely been carried on aircraft the recoil tends to damage the aircraft.

There was some work for a few decades after World War I on a way around the problem, in the form of the "countershot" guns. A countershot gun was a cannon that fired a shell out the front while firing a counterweight out the back.

The "countershot" was generally a cardboard cylinder of buckshot packed in grease that could survive being shot out the back of the cannon but would break up quickly, preventing it from being a major threat to anyone or anything behind the aircraft. Nobody ever used such weapons to any extent, and they remain something of a historical curiosity.

Countershot Cannon

- As an interesting footnote to the discussion of momentum, here's an extremely devious puzzle. Suppose that we have a fixture consisting of a short pipe standing up on a table. The top of the pipe is fitted with a hub that is free to spin around in the horizontal plane. The hub has an L-shaped hollow tube sticking out each side. An air blower is connected to the pipe below the table; when the blower is turned on, air is forced up through the pipe, through the hub, and out the two tubes, forcing the hub to spin.

Now suppose that the blower can be put in reverse, so that it draws in air through the fixture instead of blowing it out. What happens to the hub? Does it spin as it did before, or does it spin in reverse?

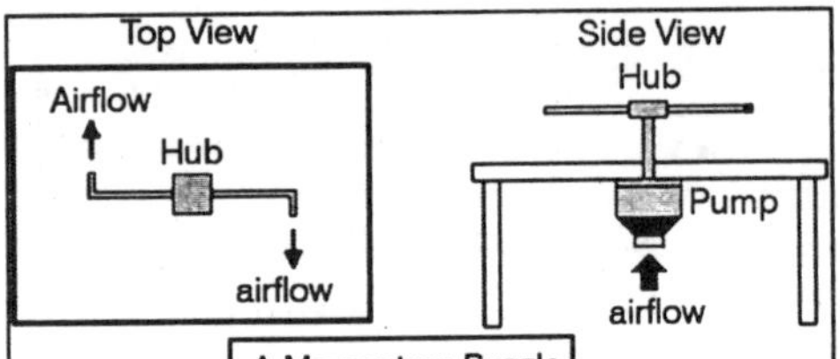

A Momentum Puzzle

As it turns out, it doesn't spin at all. The air drawn in from the ends of the tubes is static, only acquiring momentum as it travels inward. Conservation of momentum means that the hub will remain stationary.

Chapter 2

Forces in the Universe

MECHANICS IN THE REAL WORLD

- The examples used so far in this document have generally ignored friction, which makes the math much simpler, but in the real world friction can't be ignored. The friction of an object falling through the air, its drag, is not significant for a compact object falling a short distance, but otherwise that drag ultimately limits how fast the object falls to a more or less fixed "terminal velocity".

The actual terminal velocity depends on the shape and smoothness of the object. A streamlined object has less drag than a blunt one, and a smooth, polished surface has less drag than a rough one. This is because the air consists of a mix of various kinds of molecules.

On the average, except for the occasional breeze or wind, these molecules are not moving at any appreciable velocity. When a moving object hits the molecules in its path, it transfers momentum to them that causes them to bounce away.

Suppose a cylinder is shot out of a cannon, end first. When the flat end surface collides with the molecules, they bounce away in the direction of the cylinder's motion, resulting in a maximum change in their momentum. Since force is the change in momentum, this means a maximum force to slow down the cylinder.

This also implies that the force slowing down the cylinder is proportional to its frontal cross-sectional area, since that increases the number of molecular collisions. A long slender cylinder fired out of a cannon has less drag than a short fat cylinder of the same mass. Now suppose the cylinder is machined to look like a dart with a sharp point and tapered sides. As it flies through the molecules, the tapered sides tend to cause the molecules to bounce off to the sides. Relatively little momentum is transferred in the forward direction to the molecules, while the momentum transferred into the cylinder by the sideways motion of the molecules is, on the average, the same on all sides and balances out. The result is that the dart has much less drag than the cylinder. A rough surface similarly has faces at a microscopic level that can cause the molecules to bounce forward; smoothing and polishing the surface of the dart reduces these losses. There are actually a number of other, subtler considerations in designing an object to reduce drag, but they are well beyond the scope of this document.

These same principles apply to falling objects. Of course, a parachute is an application of such principles; an open parachute has much more drag and a much lower terminal velocity than parachute in its pack, since the open parachute has much more cross-sectional area to its line of motion for the same weight.

No matter how smooth or streamlined an object is, drag forces increase with speed. Since momentum is directly proportional to velocity, it might seem that drag forces increase linearly with speed, but that's only true for small objects moving at low speed; due to a complicated set of other effects, such as compression of and turbulence in the medium, the drag forces for small objects at high speed or large objects at almost any speed roughly increase with the square of the velocity. In any case, at some speed, the force of drag matches the force of gravity.

There is no longer a net force on the object, and the object cannot accelerate to a higher speed. It has reached its terminal velocity.

Of course, this does not apply to objects falling in a vacuum. This needs to be emphasized because we live at the bottom of an atmosphere, and it sets our preconceptions on how things fall. If the Earth had no atmosphere, a gram of feathers would fall at the same rate as a gram of steel. That's not too hard to swallow, but a gram of feathers would also fall at the same rate as a tonne of feathers, or a tonne of steel.

In fact, everything would fall to earth at the same rate, increasing in speed by 9.81 meters per second every second until impact, no matter how heavy it was or what its shape was. Gravitational force depends on the product of the masses of two objects involved: double the mass of one object, it doubles the force, but by Newton's second law $F = M \times A$, the doubled mass means that twice the force gives the same acceleration. Heavy objects do not fall faster than light ones in a vacuum.

It is drag that makes this simple fact somewhat counterintuitive. A feather is like a parachute, having a large cross-sectional area relative to its weight; if feathers were crushed together into a mass, of course they would fall much faster. The issue of drag did much to confuse scientists for millennia.

It was Galileo who managed to cut through the confusion; his decisive work on the subject involved rolling heavy and light balls down ramps, where the rate of fall was slow enough to allow him to accurately measure the time it took for each ball to reach bottom with the crude instruments he had available to him. The truth of the story of his dropping two balls off the soon-to-be-leaning Tower of Pisa is generally regarded as a myth.

It should be noted here that one of the other laws of unrealistic cartoon physics is that, if Wile E. Coyote falls off a cliff alongside, say, an anvil, the anvil will somehow move to a position directly above him before impact.

(Specialists in cartoon physics summarize this principle as: "Everything falls faster than an anvil.") In the real world nature is indifferent, sometimes to the point of seeming perverse -- a notion described in the well-known principle of "Murphy's Law", or "everything that can go wrong, will" – but never deliberately sadistic.

- Similarly, if Dexter tries to push along an object sitting on the ground, he has to deal with a force of friction proportional to the object's weight and to the roughness between the two surfaces. Interestingly, though it's obvious from experience once pointed out, the force of friction is greater to get a motionless object to move than to keep it moving.
- The friction force that resists putting an object in motion

is known as "static friction", and the lower friction force that resists keeping an object in motion is known as "sliding friction" or "kinetic friction".

Of course, the friction force can be reduced by making the sliding surfaces smoother and by using a lubricant. Using rollers or a dolly also greatly reduces the force of friction in moving the object, though there are still smaller "rolling friction" forces at work. One of the best ways to overcome friction is to support an object on a blanket of blown air, though this requires a certain amount of relatively complicated technology and a power source.

ELEMENTARY MECHANICS AND MACHINES

- We now know enough to consider the operation of simple machines, such as levers, ramps, gears, pulleys, screw jacks, and so on. An understanding of these devices is based on elementary concepts of force, work, and power.

A lever is one of the simplest tools. Suppose Dexter wants to lift a weight or "load" using a long stout stick as a lever. He props the stick on top of a rock for support so that three-quarters of the stick is on his side of the rock, while the other quarter is on the other side of the rock and wedged under the load. The rock is said to act as a "fulcrum", by the way.

If he pulls down on the lever, his end of the stick moves three times as far down as the other end of the stick under the load moves up. The total work done on both ends of the stick is the same, but since Dexter's end of the stick moves three times farther, he only ends up applying a third of the force. This means Dexter has a "mechanical advantage (MA)" of three to one, or in more formal terms:

$$\text{Force}_1 \times \text{displacement}_1 = \text{force}_2 \times \text{displacement}_2$$

To balance a lever, the product of the mass and length of the lever on one side of the fulcrum must equal the product of the mass and length of the lever on the other side of the fulcrum:

$$\text{mass}_1 \times \text{length}_1 = \text{mass}_2 \times \text{length}_2$$

In this example, if the lever is a meter long and the fulcrum is 25 centimeters from one end, then 1 kilogram at the long end of the fulcrum balances 3 kilograms at the short end. If the fulcrum is 20 centimeters from one end, then 1 kilogram will balance 4 kilograms. Similar issues of balance comes up in other domains of physics.

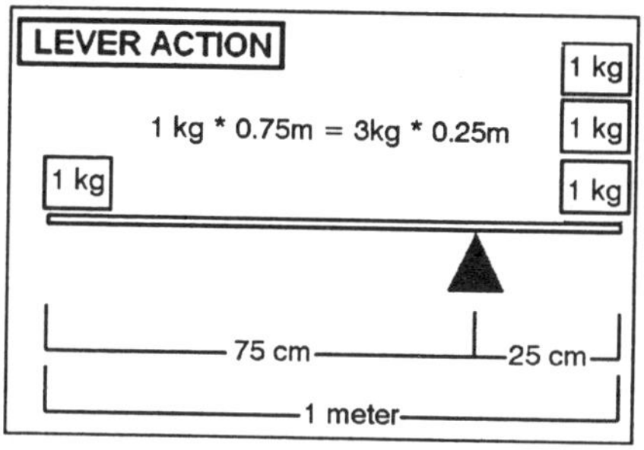

The example above involves arranging a lever to use a small force over a long distance to exert a big force over a short distance. Such a lever configuration can be reversed, using a big force over a short distance to exert a small force over a long distance; this might seem pointless at first sight, but it is useful in building a catapult.

The Medieval catapult known as a "trebuchet" (pronounced "tray-byoo-shay" though sometimes Anglicized to "tray-bucket") consists of a heavy weight connected to the short lever arm and a sling for a stone projectile connected to the long lever arm. When released, the heavy weight moves the short lever arm over an arc, with the long lever arm moving over the same arc but covering a longer distance, meaning the sling moves the payload move at a higher velocity.

The increase in velocity is the inverse of the mechanical advantage: if the long lever arm is, say, five times longer than the short lever arm, the sling and its payload has five times the velocity of the short lever arm.

- This configuration of lever, incidentally, is known as a "first class lever", meaning that the load is on one end, the work is exerted on the other end, and the fulcrum is between the two. There are two other lever configurations. The "second class lever" features the fulcrum on one end, the work on the other end, and the load between the two: it's a good scheme for nutcrackers, manual paper punches, and the little wall-mounted press I use to crush aluminum cans for recycling.

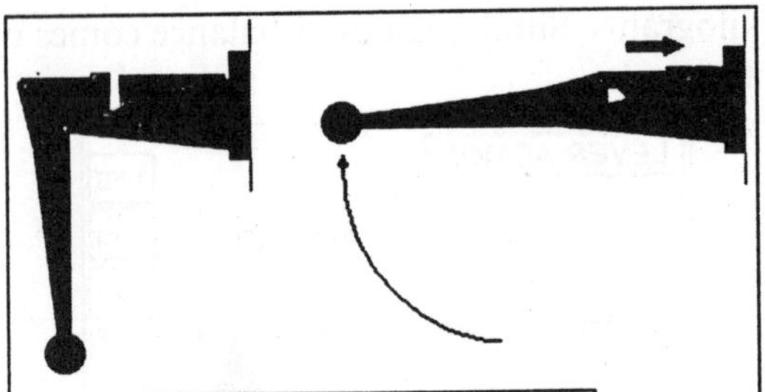

Second Order Lever: A Can Press

The concept of mechanical advantage applies to the second class lever basically as it does to a first class lever. If Dexter has a nutcracker with the jaws clamping down on a walnut a third of the way from the hinge to the grip point, he still gets a mechanical advantage of three to one. However, a "third class lever" which features the fulcrum on one end, the load on the other end, and the work exerted in between always has a mechanical advantage of less than 1, meaning the force applied to the load is always less than force exerted on the lever. In reality, a third class lever is not used to obtain a mechanical advantage as such. It is used to extend reach, for example in ice or sugar cube or salad tongs. It can also be used as a type of catapult mechanism: if a force is applied near the base of a pivoting arm, the tip of the arm will move over a longer path in the same amount of time, allowing it to throw off a load at high velocity.

Third-Order Lever (GVG/PD)

A ramp is another simple machine. Suppose Dexter wants to lift a load one meter into the back of a truck. Lifting it straight up is a good way to hurt himself, so he finds a stout plank three meters long and rolls the load up the plank on a dolly. Once more he only does a third of the work for three times the distance, and has a mechanical advantage of three to one.

A screw jack is simply a ramp wrapped around an axle, with the axle rotated by a handle used as a lever. If the circular motion of the handle is 30 centimeters per revolution, for example, and one turn of the screw lifts a weight by one centimeter, then the mechanical advantage is 30 to 1.

My little Toyota car has a scissors-type screw jack, in which rotating the handle turns a horizontal screw that raises a mechanical scissors arrangement. Obviously a detailed analysis of this device is a little more complicated, but a simple "input-output" analysis shows that if one rotation of the handle travels 30 centimeters to raise the jack one centimeter, I have a mechanical advantage of 30 to 1.

There are other ways to obtain "leverage". For example, a windlass consists of a crank with a radius Rc turning an axle with a radius Ra to wind up a rope. One turn of the crank means one turn of the axle, and so the mechanical advantage is Rc / Ra.

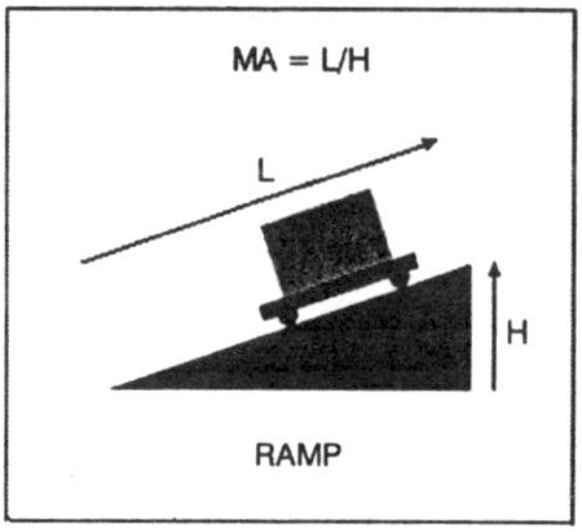

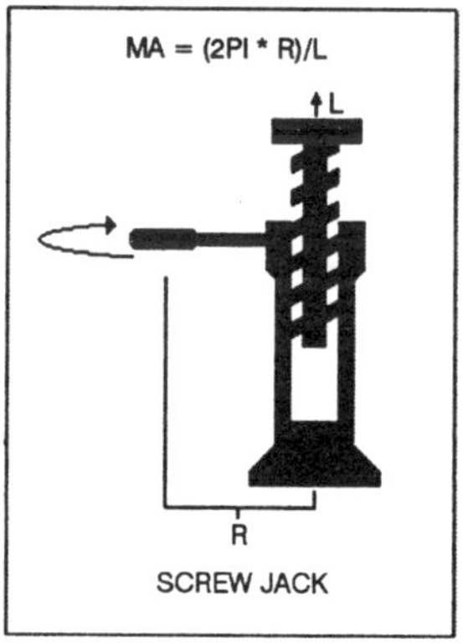

A simple pulley doesn't provide any mechanical advantage, it merely allows a force to be applied in a more convenient direction. For example, a rope slung over a pulley might be tied to the hitch of a car, with the car then driving forward to lift a weight off the ground. A block and tackle involves a pair of pulleys looped together. Pulling on the rope out of one end of the loop makes the loop shorter, but as the loop has two sides, it gets shorter at only half the rate the rope is pulled out of the loop. This means a mechanical advantage of two to one. If Dexter makes a block and tackle with two loops, he gets a mechanical advantage of four to one; if he makes it with three loops, and he gets an MA of six to one. Gears are probably the best-known of all simple machines. Two gears meshing together are basically just a set of wheels moving against each other, with the teeth ensuring they don't slip. If the driving gear is only a fifth as big in diameter as the gear it is driving, then the driving gear performs five rotations for every rotation of the driven gear. or has a "gear ratio" of 5:1. This also means an MA of 5:1.

A classic manual or "stick" automotive transmission is no more than a box full of gears of different ratios. For cruising on level ground, a high gear ratio is used, providing high rolling speed with a minimum of power. For climbing up a hill, a low gear ratio is used, providing low rolling speed with a maximum of power. Of course the transmission has a gear scheme that allows putting the wheels in reverse.

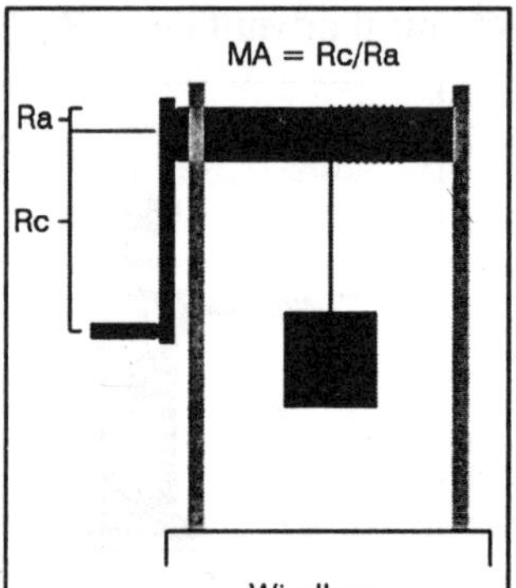

Windlass

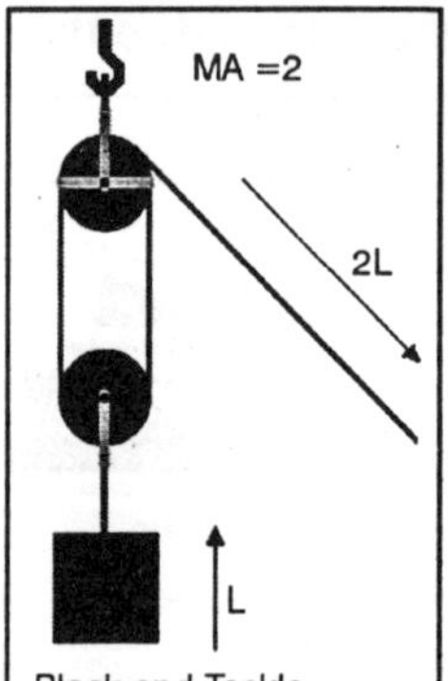

Block and Tackle

The transmission is linked to the engine through a spring-loaded friction plate or "clutch". If Dexter is wants to change the gear ratio, he steps on a clutch pedal to release the clutch plate from the engine driveshaft, which temporarily "unloads" the transmission. He then changes the gear ratio with the stick and releases the clutch pedal, loading the transmission again.

- Traditionally, manual transmissions have four to six gear ratios. Now many cars have a "continuously variable transmission (CVT)" that can vary continuously from its lowest gear ratio to its highest gear ratio. Actually, calling it a "gear ratio" is somewhat misleading, since a CVT doesn't have gears as such.

To get the basic idea of a CVT, imagine a small car with a transverse (sideways) mounted engine and front-wheel drive. There's a disk on the engine shaft and a disk on the front-wheel driveshaft, with the two disks linked only by a belt. Obviously, if there's a disk with a small diameter on the engine shaft and a disk with a big diameter on the driveshaft, the gear ratio is low, good for climbing hills; if the engine shaft disk is big and the driveshaft disk small, then the gear ratio is high, good for economical fast cruise.

The magic trick in a CVT is to make the two disks effectively change diameter. This is done by actually making each in the form of *two* disks, in the shape of shallow cones with the tips together, one of the cones being fixed to the shaft and the other able to be moved up and down the shaft using an actuator system. Pulling the two cones apart makes the belt ride lower in the space between the cones, reducing the diameter over which the belt is driven, while drawing the cones together makes the belt ride higher, increasing the diameter.

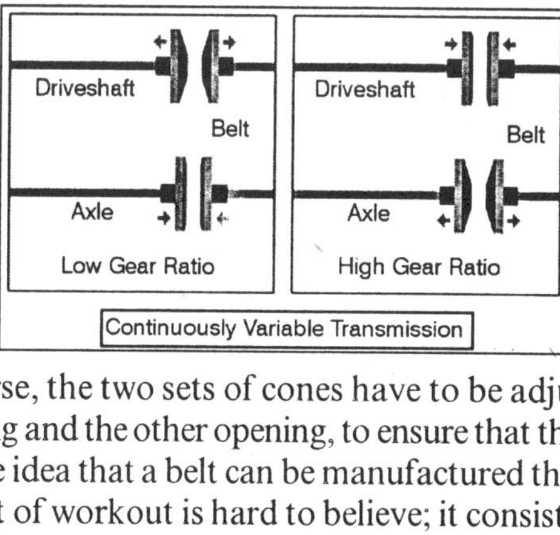

Continuously Variable Transmission

Of course, the two sets of cones have to be adjusted in step, one set closing and the other opening, to ensure that the belt doesn't go slack. The idea that a belt can be manufactured that will put up with this sort of workout is hard to believe; it consists of dual sets of steel bands, held in place with a string of strong metal cross members. The only shift settings necessary for a CVT are reverse, park, and forward. A CVT works best at relatively low speeds and in stop-and-go traffic. Due to friction losses in the belt, it offers no particular benefits at high speeds. The basic concept of the CVT is not new, going back to the Renaissance polymath Leonardo da Vinci at least. The first practical CVT design was patented by Dutch inventors Hub and Wim van Dorn in 1958, and the first full production cars with CVTs appeared in the late 1980s.

FORCES AND FIELDS

- We have defined force, and discussed gravitational forces a little. Gravity is actually only one of four fundamental forces in the Universe:
- Gravitational
- Electromagnetic
- Strong nuclear
- Weak nuclear

All other forces are just indirect manifestations of these four basic forces. The only ones we actually deal with directly are the gravitational and electromagnetic forces. The strong nuclear force holds the cores of atoms together, while the weak force only shows up in certain processes of atomic disintegration, and in fact it seems so unlike a force that physicists tend to call it the "weak interaction" instead. The strong and weak nuclear forces were completely unknown to classical physicists and will not be mentioned further in this document.

The gravitational force is the most familiar of these four forces, since we deal with it all the time. Gravity is the weakest of all the four forces, and is about forty orders of magnitude (one followed by 40 zeroes) times weaker than the electromagnetic force. To visualize just how weak a force gravity is, consider that when Dexter picks up a brick, he is overcoming the gravitational attraction of the *entire Earth* on that brick. However, gravity is the only one that is effective over long distances, and so predominates over the entire Universe.

Gravity is an attractive force between two or more masses, such as the Earth and the Moon. It's not actually restricted to such large masses, since the attraction exists between any two masses of any size, but since gravity is so weak it's hard to notice in smaller masses. As Newton discovered, the size of this force is proportional to the size of the two masses, and inversely proportional to the square of the distance between them:

gravitational force (G)= constant$\times$mass$_1\times$mass$_2$ / distance^2

Where the constant has a value of 6.672×10^{-11} in standard metric units, and the distance is measured between centre of the individual masses. Double the size of one of the masses, and the force is multiplied by two. A big planet like Jupiter has much higher gravity than Earth, 2.54 times greater to be specific. Double the distance, the force is divided by four. If Dexter blasts off from Earth in a rocket, the farther he goes, the more the force falls off.

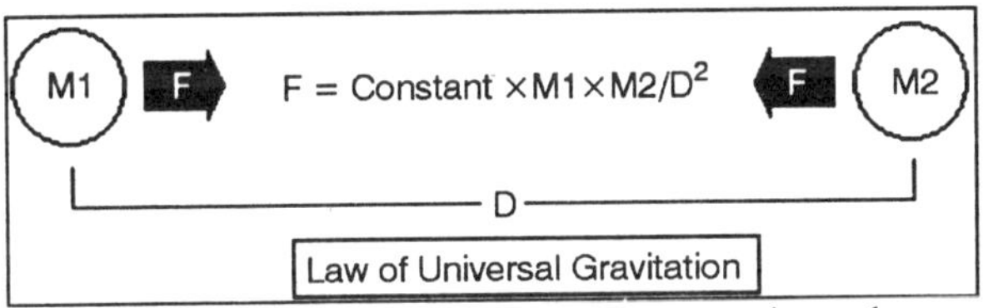

Law of Universal Gravitation

Dexter flies around the solar system to other planets all the time, experiencing different gravitational forces and having a different weight on each planet. Since the vast majority of us never get very far off the ground, much less travel to other planets, in our practical experience the force of gravity is a constant, with an acceleration of 9.81 meters per second, as discussed earlier.

It should be noted that Newton never really claimed he understood *why* gravity behaved as it did, famously stating: "I frame no hypotheses." He accepted it as simply as a given, a fundamental rule of the way the Universe worked, and though modern physics has taken the idea somewhat farther, this document accepts it as a fundamental rule as well.

- Physicists use the concept of a "field" to describe forces like gravity. This can be considered as a three-dimensional map of the direction and magnitude of the force at any point in space. The idea was originally conceived by the brilliant English researcher Michael Faraday.

Two-dimensional representations of fields usually use lines to map such fields, with the density of the lines giving the strength of the field and arrows attached to the lines to give their directions.

Leveraging off this abstraction, physicists often refer to "lines of force" or "lines of flux" in descriptions of fields and field interactions, though in classical physics the fields are actually continuous and the lines are just a convenient abstraction, much like the altitude lines around mountains and other geographical features on relief maps. The density of the lines of force gives the strength of the field and the amount of force it produces. In the special case of living on the surface of the Earth, the gravitational field is effectively uniform.

No matter where we place a mass in the three dimensions of our immediate surroundings, the direction of the force is straight down and the magnitude is effectively the same. Potential energy in this uniform field, as mentioned in the previous section, is directly proportional to the height of a mass. If the mass is raised any significant fraction of the Earth's diameter, the gravitational force falls off and calculating potential energy is trickier.

Every object in the Universe exerts a force on every other object in the Universe. However, in most analyses of gravitational forces in the greater Universe, the gravitational field is regarded as acting between simple points of matter that have no extent. This is actually a very good approximation for nearly all practical purposes.

Visualizing this from a field perspective also helps explain why gravity obeys an "inverse square" law, decreasing in strength with the square of the distance. The gravitational force of the Earth can be visualized as lines of force radiating with uniform density in all directions, with the direction of the lines pointing into the Earth.

Now let's draw an invisible shell around the Earth at some radius. All the lines of force pass through this shell. Since the surface area of a sphere is $4 \times \pi I \times radius^2$, then if we draw a second shell at double the radius, the surface area of this shell is four times that of the first.

The same number of lines of flux pass through this outer shell as they do the inner shell, but the surface area has been squared, meaning the density and so the strength of the gravitational field has been attenuated by the same factor.

- Electromagnetic forces have similarities to and

differences from gravitational forces. Gravitational forces occur between masses. Mass is irrelevant to electrical forces, they only rely on a property known as the "electrical charge" of the objects involved.

"Charge" is another fundamental concept, one that can only be explained in terms of itself: an electric field works on charges, charges are what an electrical field works on. However, charges are tangible, and can be measured. Physics defines a certain fundamental unit of charge that cannot be subdivided further. This charge is very small and is usually measured in terms of 6.24E18 of these basic charges; this amount is known as a "coulomb". Charges can flow from one object to another, with a rate of flow of one coulomb per second given as an "ampere".

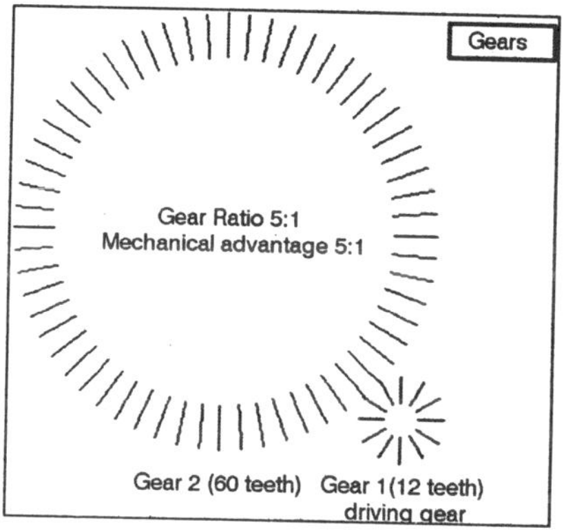

The basic law of electrical force is very similar to that of gravity. The electrical force is proportional to the product of the charges of each object and inversely proportional to the square of the distance between them:

Electrical force= constant × $\text{charge}_1 \times \text{charge}_2/\text{distance}^2$ where the constant has a value of 8.9875E9 in standard metric units. A charged object in an electric field will move in the direction of the lines of force. This relationship was discovered by the pioneering French electrical researcher Charles Augustin de Coulomb, for whom of course the coulomb of charge is named, and the electrical force is sometimes called the "Coulomb force" in his honour. Dexter can easily demonstrate this force by rubbing a balloon on his hair, a process that strips off charges; the balloon will then stick to walls or the ceiling. The same effect makes styrofoam packing peanuts difficult to handle, since they tend to stick to the fingers.

There are two types of charges, positive and negative, with objects with like charges repelling each other and objects with unlike charges attracting each other. By the way, the acceleration of an object due to an electric force on it is still proportional to its mass. The amount of charge on the object may change the amount of force on it, but given a specific force the acceleration is strictly due to the mass.

The alert reader will notice that the definition above is for "electrical force" even though the topic under discussion is "electro*magnetic* forces". The electric force is straightforward: put a charged object in an electric field, it will be accelerated towards a concentration of opposite charges or away from a concentration of like charges. The magnetic field is harder to grasp.

If we have a permanent magnet in the shape of a ring with a gap cut out of it, there will be a magnetic field between the "magnetic poles" on either side of the gap. If we throw a charged object through the gap, it will not be accelerated towards the poles; it will instead be pushed out one side or other of the gap. In more general terms, if a charged object flies across the lines of force of a magnetic field, the object is pushed *sideways* relative to its direction of motion, with a force proportional to the amount of charge on the object, the intensity of the magnetic field, and the velocity of the object.

It was recognized early on by researchers playing with electricity and magnetism that the two were related: a current passed through a wire will generate a magnetic field around it that can, for example, turn a compass needle.

In an inverse fashion, passing a wire through gap in a magnet would generate an electrical current. Further details of magnetism are a bit beyond the scope of this document and will need to be discussed elsewhere; however, it can be said without more explanation that magnetism is not a distinct force, actually being just a subtle manifestation of the same electrical force. In general, an electric field is produced by an accumulation of charges, while a magnetic field is produced by *moving* charges. Why this was so couldn't be explained by classical physics, it was just accepted as "the rules".

Incidentally, Michael Faraday arrived at the concept of a field through the classic experiment of placing a magnet under a sheet of flat sheet of paper and then sprinkling iron filings on the paper; the filings then oriented themselves along the magnetic field, allowing it to be visualized.

It was a major step forward in the comprehension of the mysterious "action at a distance" of gravity and electromagnetism, though it really didn't say much about the underlying causes of either phenomenon. Faraday lacked a formal education, having dragged himself up from dire poverty at great effort, but he had one of the most well-developed senses of physical intuition in the history of the science.

- As mentioned, electromagnetic forces are vastly more powerful than gravitational forces. If Dexter picks up two bricks and holds them out in parallel in front of him, he will not be able to sense any gravitational force between them, and in fact he wouldn't be able to sense any force with anything less than the most exacting experimental apparatus.

  If he picks up two magnets, each the size of a brick, and holds them in front of him, they will slam together and he won't have the strength to pull them apart.

Both electromagnetism and gravity are long-range forces, and so it seems odd that gravity, which is so vastly weaker than electromagnetism, is the predominant force in the Universe at large scales.

There is a major difference between electromagnetism and gravity, however: as mentioned, electrical charges have "polarity" they can be positive or negative, with a repulsive force for like charges and an attractive force for unlike charges. In contrast, gravity is, as far as anyone has ever seen, is always attractive.

The strength of the electromagnetic force also limits its long-range effects. It is so strong that it tends to pull charges from one object to another. If Dexter scuffs his feet on a carpet to get a charge imbalance and then touches Deedee, charges jump between the two in a spark, correct the imbalance, and Deedee says: "Ouch!". Similarly, if on a larger scale, on a hot day rising air masses in thunderhead clouds will create a charge imbalance that will be corrected with a lightning bolt. Charge imbalances tend to be self-correcting, and overall the Universe remains electrically neutral.

Since gravity operates on a cosmic scale, the fact that it is very weak is highly significant: if it were substantially stronger, the entire Universe would collapse in on itself, and that would be the end of that. However, it must be emphasized that, except for gravity, the forces we see in our daily lives from our own physical work, the operation of a car or similar machine are derived from the electrical force, if sometimes very indirectly.

It should be noted that in recent years hints have arisen that there may be some sort of repulsive gravitational force. It was learned in the 20th century that the Universe is expanding, having been born in a highly dense and energetic state in a "Big Bang" billions of years ago, then expanding and cooling off to form planets, stars, and galaxies.

The discovery that the Universe seems to be expanding at a faster and faster rate has suggested the existence of a mysterious "dark energy" that produces a negative gravity effect. However, nobody's really sure what dark energy is, and it's far beyond the scope of this document anyway.

- As a footnote to fields, inventors of perpetual-motion machines and the like seem to have an odd attraction to fields. They particularly like to play with magnets. It doesn't take much inspection to show that fields violate neither energy conservation nor Newton's laws of motion.

As far as energy conservation goes, an object in a field can acquire no more energy than is put into it by work. The simplest example is raising a mass in a constant gravitation field. As described in the previous chapter, on Earth this requires an amount of work given by:9.81 × height

which is of course the potential energy acquired by the mass. The same reasoning applies to electric and magnetic fields, the only main difference being that it is much easier to manipulate or create these fields, for example by moving permanent magnets around or generating a magnetic field with a current. Moving the magnets and producing a current requires work, however, and there's no way to get more work out of any such arrangement than is put into it.

As far as Newton's laws go if Dexter drops an object on Earth, it is technically not as much true that it falls to the ground so much that the object and the Earth fall towards each other, each having the same but opposite momentum in the interaction. Since the Earth's mass is presumably vastly greater than that of the object, the velocity of the Earth is immeasurably smaller.

If it were a case of two planets of equal size, they would fall towards each other, accelerating at the same speed, and meet in the middle. The total momentum of the system at all times in any case is zero.

Another misunderstanding about fields is that they consume energy, even in static scenarios. This isn't true: although the Earth's gravity is pulling down on Wile E. Coyote's anvil at all times while it is hanging from a cliff, simply pulling on it doesn't use up any energy. The Coyote had to expend energy to get it to the top of the cliff, and it will release energy when it falls on him, but as long as it remains stationary, not the slightest amount of energy is expended.

Of course, what confuses the matter is that if the Coyote falls off the cliff and grabs on to a branch, he'll expend energy just hanging there. In reality, if a plastic dummy of the Coyote ballasted to have the same weight were tied to the branch, it would stay there indefinitely with no consumption of energy. The Coyote, like any living being, expends energy in biological processes for as long as he is alive and has to expend additional energy to perform any action, such as keeping his fingers wrapped around the tree branch.

From an energetic point of view, the Coyote is something like the kind of inflatable yard displays set up on holidays ghosts and giant spiders at Halloween, santas and snowmen at Christmas, monster bunnies at Easter. If they were airtight, they could be inflated and left in place with no additional input of work, but they aren't and their structure must be maintained by the continuous operation (and energy drain) of a small fan.

If the energy supply to the fan is cut off, the display collapses, and if the Coyote doesn't have periodic inputs of energy, he collapses as well. This is why he is so desperate to catch that Roadrunner.

There is something mysterious about fields, and since they seem somewhat magical it is tempting to think that they are even more magical than they are.

Despite that, from a fundamental physics point of view a gravitational field pulling together two masses is not all that different from, say, a stretched spring pulling together two masses. It's just that with a gravitational field the "spring" is invisible.

Chapter 3

Rotational Motion

The previous chapters focused mostly on motion in a straight line. The same principles apply to rotational motion, but end up being reinterpreted in some interesting ways. This chapter provides a short study of the physics of rotational motion.

ROTATIONAL MOTION / ANGULAR VELOCITY

To consider rotational motion, let's imagine that Dexter has built a test apparatus, consisting of a mass linked to a rod that can be pivoted around a pole. The mass is very compact, so its size and shape can be disregarded. The mass of the rod is small and can be ignored.

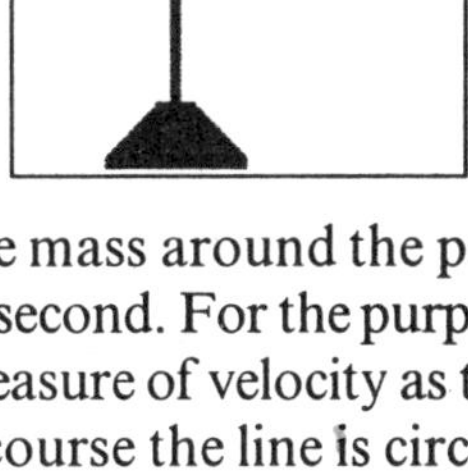

The rod spins the mass around the pole in a circle at a rate of, say, six meters per second. For the purposes of this discussion, we can refer to this measure of velocity as the "linear velocity" of the mass, though of course the line is circular, not straight. We define this as linear velocity in order to introduce a different concept of velocity applicable to rotational motion, known as "angular velocity". If the mass is positioned on the rod one meter from the pole, then the circular path length of the mass is one meter times 2π (two times pi), or about 6.28 meters. That means that the number of revolutions the mass makes around the pole in one second is: velocity of mass / ($2\pi \times$ length of rod) = 0.955 revolutions per second.

This defines the angular velocity of the mass as 0.955 revolutions per second. In general, we can define angular velocity as follows: angular velocity = linear velocity / ($2\pi \times$ radius) with angular velocity given in revolutions per second. Notice the clear difference between linear velocity and angular velocity:

- Linear velocity gives the number of meters the mass travels in a second, which means that for a given value of linear velocity the actual number of revolutions per second drops as the length of the radius, and the length of the circular path around the rod, increases.
- Angular velocity gives the number of revolutions of the mass in a second, regardless of how big a circle it travels around, meaning that for a given angular velocity, the linear velocity increases as the radius increases.

Suppose that the mass can be slid forward and back along the rod, and that it Dexter moves it in so that it's only half a meter from the pole. Suppose further that the rod continues to rotate at the same rate. That means that although the angular velocity is the same, 0.955 revolutions per second, the linear velocity has been cut in half, to three meters per second. Similarly, suppose Dexter moves the mass out along the rod until it's two meters from the pole, and once again the rod continues to rotate at the same rate. The angular velocity is still the same, but the linear velocity has doubled, to twelve meters per second.

- Angular velocity is not normally measured in revolutions per second. The popular unit for angular velocity is "revolutions per minute (RPM)", and of course all that needs to be done is multiply revolutions per second by 60 to get revolutions per minute. For the example above, the RPM amounts to $60 \times 0.955 = 57.3$ RPM.

However, RPM is not a very convenient unit for performing calculations. A somewhat more convenient unit for angular velocity is degrees per second, with 360 degrees in a circle. To get degrees per second, all that needs to be done is multiply the revolutions per second by 360. For the example above, the mass has an angular velocity of 360×0.955 = 344 degrees per second.

Degrees are a unit of angular measurement, and though we have defined angular velocity as the number of revolutions per second, the term "angular velocity" itself suggests that it's more properly defined as the change in angle per second. The two definitions are equivalent, with the first being more intuitive and the second being more useful for calculations.

In practice, from the point of view of calculations, there is a even more convenient unit for angular measurement than degrees, known as "radians".

One radian is just the length of one radius of a circle along the circumference of that circle, and so there are 2π, or about 6.28, radians in a circle. Degrees are in popular use and so are generally more intuitive than radians, but radians make rotational calculations much more consistent and easy to handle. For example, if angular velocity is expressed in radians, then its definition is simplified to:

Angular velocity = linear velocity / radius

The final result of calculations can be converted into RPM or degrees per second if need be, and the conversion calculations are very simple:

$$\text{RPM} = 60 \times \text{radians} / 2\pi\text{I}$$

$$\text{Degrees per second} = 360 \times \text{radians} / 2\pi\text{I}$$

Angular velocity may seem somewhat arbitrary at first sight, but a moment's thought shows how convenient it is. It is much simpler to describe the rotation.

Of a helicopter's rotor as say, 200 RPM, than it would be to describe it in terms of the linear speed of the rotor tips, as 452 kilometers per hour. Of course, if we can define a quantity such as angular velocity, we can also define the related quantities of "angular displacement", or the number of radians a revolving object has moved through; and "angular acceleration", or the rate of change of angular velocity, in radians per second per second.

CENTRE OF MASS

- Defining the concept of angular velocity, as opposed to linear velocity, suggests that other quantities useful in rotational motion can also be defined that are comparable to fundamental quantities in linear motion. In fact, there's a complete and self-consistent system of quantities for rotational motion.

The basic rules for linear motion are provided by Newton's three laws of motion, and define the basic quantities of force, mass, and acceleration. We can redefine the three laws for rotational motion by defining:

- A rotational equivalent of force called "torque"
- A rotational equivalent of mass known as "moment of inertia".
- A rotational equivalent of linear momentum known as "angular momentum".

The definitions of torque, moment of inertia, and angular momentum are intertwined, so all three definitions, and then explain them in detail in the following sections. Torque is obtained by multiplying the tangential force that sets an object spinning times the length of the lever arm, or radius, on which the force acts relative to the centre of rotation of the object:

Torque = force • radius

The moment of inertia, at least for a simple system like the mass on a rod used in the previous section, is given by:

Moment of inertia = mass × radius^2

This definition is extended for more complicated systems later. For now, just accept that this definition is the truth, just not the *whole* truth. By the way, moment of inertia is sometimes called "rotational inertia", but we won't use that term in this document. The definition of moment of inertia leads to the definition for angular momentum:

Angular momentum = moment of inertia × angular velocity

The quantities of energy, work, and power remain the same in rotational motion as they are in linear motion, though their definitions are modified as shown below.

- The "centre of rotation" mentioned above is better known as the "centre of mass". For any object that is symmetrical around its spin axis, the centre of mass is the physical centre of the object. However, if the object is not symmetrical, the centre of mass is where the object would balance if hung by a thread, using the simple rules of lever balance outlined in the previous chapter to determine the point where the forces from the two masses balance out. For example, consider an object that looks like a barbell, with a 15 kilogram mass M_1 at one end, a 3 kilogram mass M_2 at the other, and with a connecting rod of negligible mass 1.3 meters long between them.

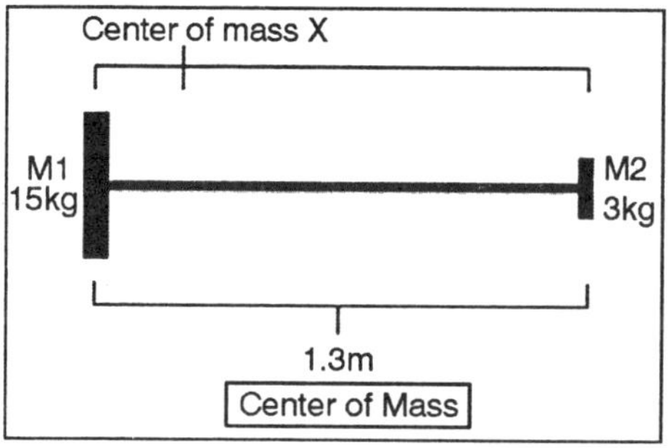

Center of Mass

Then the balance point×as measured from M_1 is given by:

$$M_1 \times X = (1.3 - \times) \times M_2$$
$$M_1 \times X + M_2 \times X = 1.`3 \times M2$$
$$X \times (M_1 + M_2) = 1.3 \times M_2$$
$$X = 1.3 \times M_2 / (M_1 + M_2) = 1.3 \times 3 / (15 + 3) = 0.22 \text{ meters.}$$

This same "balancing" approach could be extended to an assembly of masses in two or three dimensions, but although the math is not complicated, it can be long-winded, and so will not be discussed further here. Centre of mass is an important concept. For example, if Dexter is in outer space and throws a wrench so that it tumbles as it flies, its centre of gravity will travel in a perfectly straight line. If Dexter is on Earth and throws a wrench so that it tumbles around, the wrench's centre of mass will follow a smooth arc of the object's flight even as the rest of the object tumbles around that arc.

The location of the centre of mass defines the stability of an object. Suppose Dexter has a block of wood of uniform density that is, say, ten centimeters long and wide, and two centimeters tall. If the block is laid flat, standing two centimeters tall, there is no way to tip it over by pushing it on one side or another. The block lying flat is said to be in "stable equilibrium": trying to tip it *raises* the centre of gravity.

Wrench Tumbling Around its Center of Mass

Now suppose the block is stood on its side, standing ten centimeters tall. If Dexter pushes it at the top, then if the block tilts beyond an angle where the line from the centre of gravity falls outside the base of the block, it falls over. The block standing up is said to be in "unstable equilibrium": beyond the threshold of no return, pushing on it *lowers* the centre of gravity and it has no place to go but down.

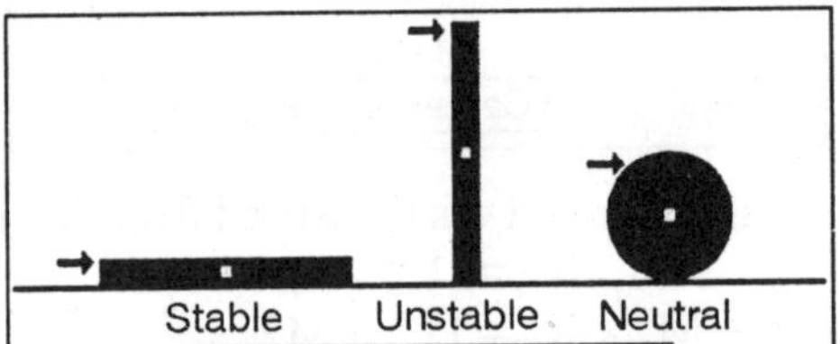

Categories of Equilibrium

This is why high-riding sports / utility vehicles are much easier to roll than low-to-the-ground sports cars. It is also possible to have "neutral equilibrium", for example in a sphere of uniform density: pushing and prodding on the sphere doesn't affect the position of the centre of gravity in the least.

TORQUE, MOMENT OF INERTIA, AND THE SECOND LAW OF MOTION

- The best way to understand torque, moment of inertia, and angular momentum is to see how Newton's three laws of motion are restated in a rotational context. It might seem logical to start with the first law of motion, which is the law of conservation of momentum, expressed with the definition for linear momentum:

linear momentum = mass × linear velocity

The definition of angular momentum is exactly comparable:
Angular momentum = moment of inertia×angular velocity

However, understanding the mechanisms of angular momentum means understanding the concept of moment of inertia as well as a few other subtleties, so an understanding of the first law in a rotational context will have to be put off for bit. That means the best first step is to consider the second law of motion. In the linear context, this is given by:

Force = mass × linear acceleration

In the rotational context, the equivalent expression is:

Torque = moment of inertia×angular acceleration

The two expressions are actually equivalent. The equivalence can be easily seen by considering the simple system described earlier, with a mass rotating at a certain radius on a rod, and the force causing the torque operating directly on the mass at that radius. Then we can restate the rotational version of the second law as follows:

Torque = moment of inertia×angular acceleration
Force×radius = mass × radius^2× angular acceleration
Force = mass × radius × angular acceleration
Force = mass × (radius × angular acceleration)
Force = mass × linear acceleration

The underlying physical principles established for rotational motion haven't changed they've just been restated in a different form.

- This shows the equivalence between the force and torque, but it doesn't really explain the concepts of torque and moment of inertia.

The concept of torque is fairly easy to understand. To get something to spin, Dexter has to push it along a direction at a *right angle* to the radius of rotation. Any force in the *direction* of that radius might get the entire rotating system to move in a straight line if it's not bolted down, but it won't contribute to rotation.

Suppose Dexter takes a rotating disk with a hex nut on the centre, and uses a wrench to get it to rotate. He applies a torque equal to the force he exerts on the wrench times the length of the wrench. Considering the discussion of levers in the previous chapter, obviously the same force will have a greater effect if the wrench is longer. It's just simple mechanical advantage, exerting a smaller force over a longer path to obtain the same amount of work. This means that the force actually exerted on the rotating disk, and so its angular acceleration, is proportional both to the length of the lever arm and the force exerted at the end of that lever arm.

One of the peculiarities of torque is that it's measured in units of force times displacement, or "newton-meters" in metric. Work is also measured in the same units. Does this imply that they're the same thing?

No, it doesn't; in fact, they're literally *sideways* to each other. Work is force on an object in the direction of its motion. Torque is exerted at a right angle to a radius, and there's no motion along the radius.

- Moment of inertia is a bit harder to understand. Let's consider our simple example again of a sliding mass on a rod, but now let's say the rod has an extension a meter long on the other side of the axis, with a handle on it that Dexter can push on to cause rotation.

Let's suppose that the mass is positioned on the rod a meter away from the axis. If Dexter pushes on the handle, the same linear force acts on the mass on the other side of the axis, and the mass accelerates just as the second law says it does.

Now Dexter moves the mass out to two meters and pushes on the handle. Since the mass moves twice as far as the handle, it only receives half the force at any one time, and its linear acceleration is only half as great as it was at one meter. This means he has to push twice as hard to achieve the same linear acceleration.

However, what we're after here is not linear acceleration, but *angular* acceleration, a change in RPM. If he doubles the radius to two meters, as noted has to push twice as hard to get the same linear acceleration, but now the circular path along which it moves is twice as long. Even given the same linear velocity in the two positions, the angular acceleration is only half of what it was for the same linear velocity when the radius was one meter.

This means that to obtain the same angular acceleration when the mass is moved from one meter to two meters, Dexter has to push *four* times as hard, or proportionally to the square of the radius. This leads back to the original definition of moment of inertia:

Moment of inertia = mass × radius^2

Once again, the same physical laws apply, they're just restated in a different form.

MOMENT OF INERTIA IN DETAIL

- As noted earlier, the initial definition of moment of inertia:

  Moment of inertia = mass × radius^2

was only for the simple mass on a rod, and was not exactly true for other cases.

Let's consider a more complicated example, where the object whose moment of inertia we want to calculate has several different masses of different size, each at a different distance from the axis of rotation of the object. This leads to a more general definition for moment of inertia:

Moment of inertia = SUM (mass×radius^2)

In simple terms, the moment of inertia of the entire object is the sum of the moments of inertia of its components. If the object had three different masses, each at a different distance from the axis of rotation, then the moment of inertia of the object is:

$$\text{Moment of inertia} = \text{mass}_1 \times \text{radius}_1^2 + \text{mass}_2 \times \text{radius}_2^2 + \text{mass}_3 \times \text{radius}_3^2$$

Let's consider more specific examples, such as a barbell with weights connected by a rod. Ignoring little details like the weight of the rod, if each weight has a mass Mi and the rod has a length L = 2×R, then its moment of inertia for rotation around the midpoint of the rod is:

$$2 \times Mi \times (L/2)^2 = 2 \times Mi \times R^2$$

If we designate the sum of the mass of the weights as Ms, then this becomes:

$$Ms \times R^2$$

If we have a pair of rods of length L = 2 × R joined in a cross, with weights with mass Mi at each end of the rods, then the moment of inertia for rotation around the centre of the cross is:

$$4 \times Mi \times R^2$$

or once again:

$$Ms \times R^2$$

In fact, this moment of inertia also applies to any thin ring or hollow thin cylinder rotating around its axis. They're exactly the same as the barbell in this respect because we still have a mass at a particular radius, and it makes no difference if it's two individual masses or a continuous ring.

Calculating moments of inertia for other objects generally requires more math than is appropriate in this document, but a short list of the moments of inertia of a few different objects of mass M will help give a better feel for the concept.

- The concept of moment of inertia makes a certain amount of intuitive sense. For example, it's much easier to rotate a barbell if the weights are close together on the bar instead of at the ends. Similarly, if Deedee is walking a tightrope, she might carry a long pole held tightly in her hands to provide balance. The longer the pole, the greater its moment of inertia, and more it prevents Deedee from tipping over.

In other cases, it's not as intuitive. Suppose Dexter has two cylindrical masses, one in the form of a hollow cylinder and the other in the form of a solid cylinder. Incidentally, they don't have to have the same diameter or the same mass.

If Dexter drops drop these two cylinders from the same height, as emphasized in the last chapter, ignoring drag they'll hit the ground at the same time.

If he were to put these two cylinders on end on top of a ramp and let them slide down, if the ramp is very slick, once again they'll reach the bottom at the same time.

Now if Dexter rolls these two objects down a plane, the solid cylinder will always beat the hollow cylinder. This is because the hollow cylinder has a greater moment of inertia, and its angular acceleration is lower for a given force. The relative sizes of the two cylinders doesn't matter any more than it does for dropping them, because the force required to get them to move is proportional to the mass, but the hollow cylinder just can't get rolling as fast as the solid cylinder.

THIRD LAW OF MOTION / WORK AND ENERGY

- After considering the second law of motion in a rotational context, it is almost too easy to consider how the third law fits in. Obviously, if Dexter generates a torque on a rotating system, the same torque is produced on him in the opposite direction.

This is most obvious in the case of a helicopter. A helicopter spins a rotor at high speed in one direction, which means that it tends to spin in the other direction. This is why most helicopters have a little rotor attached on the side of the tail, just to stop the machine from spinning; the British call this a "penny-farthing" configuration. There are other approaches to the perform the same trick. Some large cargo helicopters have rotors on both ends, spinning in different directions.

Yet another scheme is the "eggbeater" configuration, with twin rotors meshing in a side by side configuration, or the "coaxial" configuration, with one rotor sitting on top of the other but rotating in opposite directions.

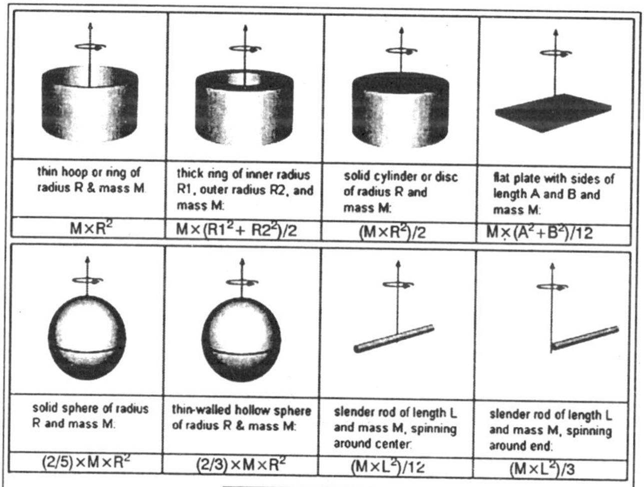

MOMENTS OF INERTIA

There is a way around the problem entirely. Some helicopters have been designed with little jet thrusters of one sort or another on the end of the rotor blades, with fuel passed up from the helicopter through the rotor blades. This causes the rotor to turn itself, and eliminates the need to cancel torque. However, for various reasons, apparently fuel consumption and noise, no "tipjet" helicopter has been produced in quantity to this time. Coaxial propellers were also used on some of the very last propeller-driven fighter aircraft. These aircraft had such powerful engines that they generated an enormous amount of torque that made the aircraft very hard to control, and so they were often fitted with "contra-rotating" propellers, with two sets of blades turning in opposite directions, driven by the same engine using a wheels-within-wheels "planetary gear" transmission.

A "contraprop" also permitted the propeller to make effective use of the high engine power without requiring a propeller of excessive diameter.

In addition, twin-engine propeller-driven aircraft often had their two propellers turning in opposite directions to cancel out the effects of torque. This wasn't always done, because it meant that the aircraft required two different types of engines, one "left handed" and the other "right handed", which complicated maintenance.

Flying an aircraft with high-powered handed engines could be tricky, since if one engine failed on takeoff the other would tend to flip the aircraft over on its back.

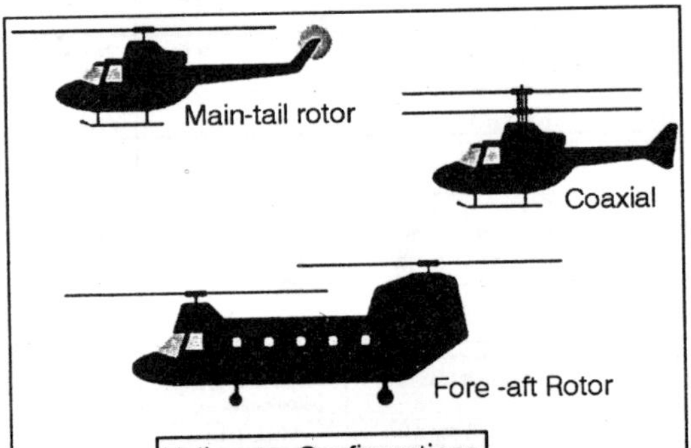

Helicopter Configurations

- It is also easy to see that the expression for work in the rotational context is similar to the expression for work in the linear context:

$$\text{Work} = \text{torque} \times \text{angular displacement}$$

and in fact the two expressions are equivalent:

$$\text{Work} = \text{torque} \times \text{angular displacement}$$

= force × radius × angular displacement
= force × linear displacement

Going back to the example of a mass on a rod rotated around a pole by a one-meter extension with a handle on the other side, then pushing the handle over an arc a meter long with a force of a newton does a joule of work. It doesn't matter how far out along the rod the mass is, the work done on the mass is the same as the work done on the handle.

The expression for kinetic energy in the rotational context is also similar to the expression for kinetic energy in the linear context: Kinetic energy = (12) × moment of inertia × angular velocity^2

and again, the two definitions are equivalent:

Kinetic energy = (1/2) × moment of inertia × angular velocity^2
= (1/2) × mass × radius^2 × angular velocity^2
= (1/2) × mass × (radius × angular velocity)^2
= (1/2) × mass × linear velocity^2

Kinetic energy is due to mass moving at a certain linear velocity. It doesn't matter if the velocity in a straight line or in a circle, the kinetic energy is the same.

CENTRIPETAL AND CENTRIFUGAL FORCE

- We now have a good grasp of the second and third laws of motion in the rotational context, and so it might seem time to go back and nail down the first law. However, we end up having to take yet another side trip before we can do that, and consider the forces on a rotating object.

As explained in the previous chapter, an object moving along a straight line at a certain velocity will continue to do so unless a force acts on it. A mass revolving in a circle is always changing direction, which means there is a force on it causing it to do so. If terms of our simple mass-on-rod example, the rod exerts a force on the mass, directed towards the pole. This is known as "centripetal force", and it is always directed towards the centre of rotation.

The concept of centripetal force sounds very much like the more popular notion of centrifugal force. Centrifugal force is what most people recognize as the force that they feel in the well-known "Tornado" carnival ride in which they are placed in a big spinning drum whose floor drops out once the passengers are pinned to the walls, or the force that squeezes the water out of clothes in washing machine during its spin dry cycles.

In fact, centrifugal force is just a parochial way of perceiving centripetal force. Riders in the Tornado feel a centrifugal force pressing them to the walls, and don't see the bigger picture of centripetal force: the spinning drum is essentially trying to throw off its passengers in a straight line, but the arc of the walls keeps forcing them back inward. During the washing machine's spin dry cycle, the water just keeps going out the holes in the basket in a straight line, while the basket pulls the clothes themselves into a spinning curve at a rapid rate.

From a casual point of view the distinction is small, but physicists get fussy about it, since using centrifugal force in the mathematical analysis of a rotating system tends to be awkward as best. It's very similar to believing that the Sun orbits around the Earth every day, which it seems to from our parochial point of view as we watch it rise and set, when the reality is that the Earth is spinning around on its axis.

However, most people don't understand the term centripetal force, and that leads to a simple rule: use the term "centrifugal force" in casual conversation and "centripetal force" in a technical context. Anyone who understands the distinction will understand; anyone who doesn't won't care and trying to explain it to them is pointless.

- One application of centripetal force is to provide artificial gravity in spacecraft. Back in the 1950s and 1960s, space stations were generally seen as big rotating wheels, spinning to create artificial gravity, as in Stanley Kubrick's movie. (This particular design actually had two wheels on a common axis and spinning in opposite directions, cancelling the angular momentum of the station as a whole and making it easier to reorient.)

2001 has come and gone and no such thing happened, but the idea remains valid. A similar but less ambitious concept is to link a pair of spacecraft by a long cable or "tether" and set them to spinning around each other. This may need to be done on a long Mars mission or other expedition where people would otherwise be in weightless condition for long periods of time. People who spend long periods of time in weightless conditions have to be retrained to walk when they return to Earth, and it would make no sense to land astronauts on Mars and then wait a few weeks for them to condition themselves until they were actually able to get around.

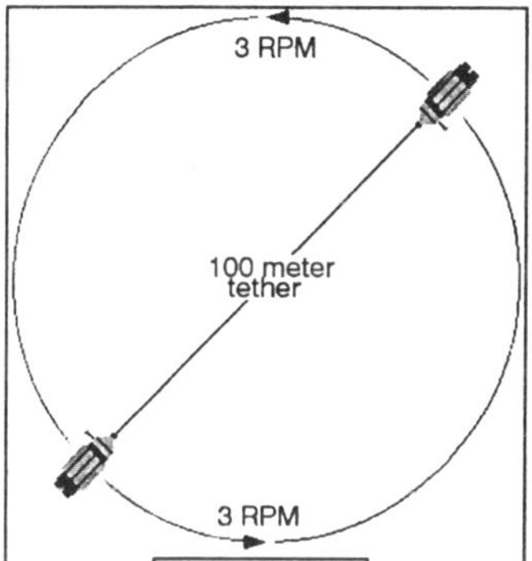

Artificial Gravity

Suppose the tether is 100 meters long, giving each spacecraft a distance from the centre of their mutual rotation of 50 meters, and the spin rate is 3 RPM, or one rotation every 20 seconds. Incidentally, the maximum rotation rate that human beings can tolerate without suffering from vertigo or nausea seems to be about 2 RPM. The formula for centripetal acceleration, given here without derivation, is:

centripetal acceleration = radius×angular velocity^2

Plugging in the quantities:

$50 \times (2\pi / 20)^2 = 4.93$ meters per second

This is about half of Earth gravity. It would be plenty for a Mars mission, since the surface gravity on Mars is only about a third of the Earth's.

Since setting two spaceships in motion on a tether complicates maneuvering, another scheme has been proposed, in which a spacecraft carries a centrifuge to condition the astronauts for a session each day. Incidentally, Earth-based centrifuges were used in test and training for astronauts in the early days of space flights, and one of the results of this effort was to characterize just how many gees of acceleration a human being could stand.

An astronaut would ride in a capsule on a couch designed to conform to his body, with his back to the outside of the arc; above about six gees he wouldn't be able to lift his arms out of the couch. Above 16 gees it was almost impossible to keep from blacking out, though one researcher did 20 gees for a short time on a dare. Astronauts would have their backs mottled from broken blood vessels after a ride on the centrifuge.

The centrifuge used by the Americans also had the interesting feature that the capsule could be flipped around so the astronaut would then be *facing* the outside of the arc; they called this routine "eyeballs in, eyeballs out". An astronaut's head had to be strapped to the couch so he wouldn't break his neck when the capsule flipped over. The maximum limit for such a transition was about nine gees.

CONSERVATION OF ANGULAR MOMENTUM

- The physics of rotational motion also introduces a number of elaborations to the physics we learned for linear motion. The basic physical rules stay the same, but new games are played.

For example, let's return to our moving mass on a rod, rotating around a pole. In the previous example, Dexter moved the mass in and out, keeping it at the same angular velocity. In reality, the mass wouldn't stay at the same angular velocity as it moved in or out, unless the rod was being driven from the pole or its tip.

Let's see what actually happens if Dexter moves the mass in or out in or out, and the mass is not being driven. If the mass is moving at a linear velocity of six meters per second when it is one meter from the pole, and it is moved to a half meter from the pole, assuming no friction or other forces on the rod or mass, the law of conservation of momentum ensures that the mass will continue to move at six meters per second.

Since the radius of the circle has been cut in half, so has its circumference, and the mass now revolves around the pole at twice the angular rate that it did when it was a meter from the pole. Moving the mass inward causes the angular velocity to increase.

Similarly, if the mass moves out along the rod until it is two meters from the pole, in the absence of any external forces it is still moving at six meters per second, but the circle is twice as big. The angular velocity is now cut in half. Moving the mass outward causes the angular velocity to decrease.

- This is an example of the principle of the "conservation of angular momentum", or Newton's first law of motion applied to rotational motion. There's no new fundamental principle involved, it's just the implications of the conservation of linear momentum applied to a rotational system, but using the same basic rules in a different games gives new results.

There are many practical examples of the law of conservation of angular momentum. A very popular example is to visualize Deedee on ice skates going into a spin. As she goes into the spin, she leaves her arms spread out wide, and the spin is slow, but as she draws her arms in to herself, she spins faster and faster. When she spreads her arms out wide again, the spin slows down.

A more cosmic example is the formation of a "neutron star". A neutron star is a very small but dense star, the size of a mountain but as massive as a star like the Sun. It is created as a "cinder" left over from a "supernova", an explosive collapse of a star much bigger than the Sun. The supernova explosion blasts off the outer layers of the star, while the star's interior falls inward to create the neutron star.

As the interior falls inward, conservation of angular momentum causes the rate of rotation to increase, and after the explosion the neutron star is spinning very fast, much faster than the oversized star that gave it birth. Young neutron stars often have "hot spots" that generate radio noise, and so they emit radio pulses as they spin, bringing the hot spot in and out of view. Such a "pulsing" neutron star is known as a "pulsar". A spinning neutron star loses angular momentum through indirect forces over time, slowing down its spin. Astronomers can roughly estimate the age of a pulsar by its spin.

In the case of Deedee going into a spin on her skates, or a star collapsing into a neutron star, there's an existing spin rate that is increased by reducing the radius. Another aspect of conservation of angular momentum is that it obeys the law of action and reaction: angular momentum produced in one direction has to be balanced by angular momentum in the other so the net sum is still zero.

Dexter doesn't have much trouble twisting around while he's standing on the ground, since the massive Earth easily soaks up the reaction, but if he's trying to twist around when his feet are stuck to a small freely floating or rotating platform, it's not so easy: if he shifts his upper body in one direction of rotation, the platform will twist around in the other direction.

When astronauts first tried to perform work in space outside of a spacecraft, they found that conservation of angular momentum made life very difficult for them. If they tried to turn a bolt one way, they would spin around in the other. This was one of those things that should have been obvious in hindsight, but nobody had quite thought things through ahead of time.

The solution was straightforward: fit a spacecraft with handholds, footholds, railings, and the like to allow the astronaut to have something to hang on to. One particularly ingenious scheme was later introduced with the US space shuttle, which has a long robot arm. The arm could be fitted with a foothold platform that an astronaut could stand on, with an operator moving the astronaut around as needed as an extension of the arm.

- Another example of the conservation of angular momentum is known as the "Coriolis effect". If Dexter stands at the North Pole, properly dressed for the weather, and points a finger straight out in front of him, his fingertip will trace out a full circle in 24 hours. Now he goes to the equator, discards his parka in favour of a Hawaiian shirt, and hold his finger straight up over his head, it will also trace out a full circle in 24 hours. However, at the Pole, the actual distance traveled by his finger in 24 hours is a little over six meters, while at the equator it's about:

  $\pi \times 12{,}750 = 40{,}055$ kilometers

The angular velocity is the same in both places, but the actual linear speed along the circular path is very much greater at the equator, since the distance between Dexter and the Earth's axis of rotation is very much greater as well. Similarly, suppose Dexter fires a cannon at a high northern latitude with the gun pointed due south, so that the cannonball arcs high through the sky towards the equator. The cannon is moving from west to east along with the rotation of the Earth, and of course the cannonball has the same sideways motion once it's fired. This isn't noticeable at short range, since everything in the vicinity is moving from west to east at effectively the same rate. Farther south, however, the distance of the cannonball from the Earth's axis of rotation will increase, and the linear velocity of the Earth's surface underneath the cannonball will increase accordingly.

The cannonball will start to lag behind the Earth's rotation, curving to the west. This is the Coriolis effect, named after the French physicist Gaspard de Coriolis. It has to be calculated for missile and spacecraft launches, and is particularly important in weather studies, since it has a definite effect on air or cloud masses moving north or south. One of the interesting subtle results of the Coriolis force is that large cyclonic storms rotate counterclockwise in the northern hemisphere and clockwise in the southern hemisphere.

There is also a folk tale that the whirlpools formed when sinks and bathtubs are drained have different spins in each hemisphere, but Coriolis forces are too subtle to have any measurable effect at such a small scale, and demonstrations that claim to show otherwise inevitably turn out to be rigged.

- The discussion of rotational motion so far has been as simplified as possible. While it's about as much as anyone needs to know if they're not going on to engineering studies, it's worthwhile to point out some of the complications to prevent things from appearing simpler than they really are.

One of the nonintuitive aspects of angular momentum is that is it a vector quantity, just as linear momentum is. The tricky question is: what is the direction of the angular momentum vector? If a mass is spinning on a cord, then its direction of its motion changes continuously, covering every possible direction in its plane of revolution. The only direction left is the axis of rotation.

There's the second question of whether the direction of

angular momentum of a object spinning in the horizontal plane is up or down, but the convention followed is a "right-hand rule". This means that Dexter grasps a rotating shaft with his right hand so that his fingers point in the direction of the rotation, then sticks his thumb out, his thumb gives the direction of the angular momentum vector. The direction of the angular momentum vector of a platter spinning counterclockwise on a table is up. Spin the platter clockwise, and the direction is down.

One of the most popular demonstrations of the vector nature of conservation of angular momentum is an experiment often performed in basic physics classes. Deedee stands on a platform that is free to rotate and holds a bicycle tire with handgrips on an axle in front of her, with the tire in the vertical plane. Dexter uses a power tool of some sort to set the tire spinning rapidly.

As long as Deedee holds the tire vertical, nothing happens. If she tries to tip it to one side, she starts to spin on the platform in the opposite direction to the rotation of the tire. If she manages to get the tire completely horizontal, she and the platform spin at an angular velocity proportional to the ratio of the moment of inertia of the tire and the moment of inertia of Deedee and the platform.

- The vector conservation of angular momentum is also illustrated by a spinning top or gyroscope, which tends to remain upright no matter where it is moved.

Gyroscopes have long been used for autopilots and "inertial navigation" systems, since they provided a reference to establish the motion of an aircraft or other vehicle. Since friction on the gyroscope bearings could affect gyroscope behaviour, resulting in a "drift" between the indicated and actual positions, various schemes were used to ensure the least friction possible, including liquid and gaseous bearings.

Gyroscopes have been increasingly replaced in such applications by more sophisticated systems using light beams focused in a loop, or sets of micro-machined acceleration sensors. Details of such technologies are a topic for another document.

Gyroscopic action is often used as a stabilization technique. Firearms and artillery generally have "rifling", or spiral grooves in the barrel, to set a projectile spinning when fired. This also helps average out the unbalanced effects of wind resistance due to imperfections in the shape of the projectile. Similarly, spacecraft generally have to maintain a specific orientation in space to perform their mission, and so them are often set to spinning to stabilize them.

Such "spin stabilized" spacecraft will have a communications dish antenna mounted on the spin axis and point the spin axis back to Earth so that constant communications can be maintained.

However, they also may need to keep some of their equipment, such as telescopes or communications dishes, pointed at a specific target, and so such gear is mounted on a section of the spacecraft that is mechanically spun to cancel out the effects of the spacecraft's spin. Electrically connecting this confusingly-named "despun" section to the rest of the spacecraft is a tricky matter, requiring devices known as "rotary couplers", but that's another story.

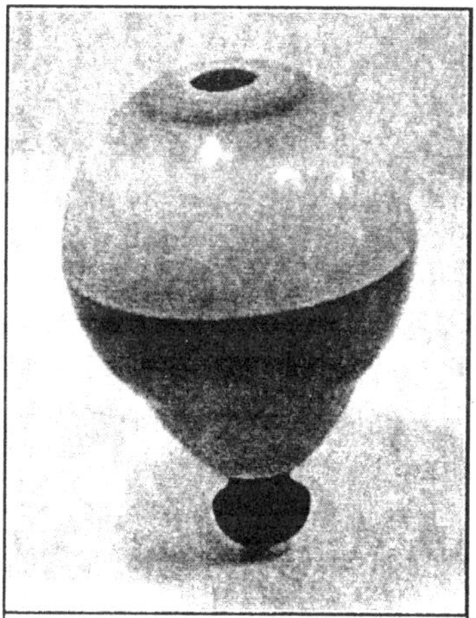

Spinning Top (GVG/PD)

Most modern spacecraft use what is known as "three axis stabilization" to maintain orientation without spinning. Although three axis stabilized spacecraft use little rocket thrusters to maintain their orientation, they may also use a set of "reaction flywheels" set in the X, Y, and Z directions to maintain orientation, adjusting the speed of a specific flywheel to transfer angular momentum and change the spacecraft's orientation slightly. Tops and gyroscopes illustrate one of the most counterintuitive features of vector conservation of angular momentum. Any force that tends to tip the top or gyro to one side sets up a torque that actually moves the top or gyro at a right angle. This is why a top starts going in lazy circles as it burns off its rotational energy through friction. Gravity is trying to pull it over, but the result is to cause the top to "precess" in circles that get wider and wider until the top falls over. There also may be a bit of instability in the precession, with the top performing a wobble or "nutation" as it spins around.

An even more common example is the tendency of a coin to roll away when dropped on a smooth surface instead of simply dropping flat. If the coin has any forward motion when it lands on its edge instead of its face, it tends to roll. That creates angular momentum in the coin that resists the tendency of gravity to pull it over, and it keeps right on rolling.

If it doesn't run into something first, it will eventually start to tip and then turn into a tightening spiral that eventually bring the coin flat to rest. One amusing experiment along such lines is to shake a coin in a transparent plastic food tub, like those used to sell caramel popcorn or the like. It turns out to be easy to get the coin to start rolling around the sides of the tub in a whirring spin, and if the coin is heavy enough, slightly difficult to get it to stop.

FOOTNOTE: MACH'S PRINCIPLE

- Having gone through the basic concepts of linear and rotational mechanics, it is useful to introduce a subtlety that would have been too confusing to introduce earlier.

The easiest way to consider this subtlety is to perform a little experiment in rotational dynamics. Suppose Dexter hangs a bucket of water on the end of a rope, and then twists the rope so the bucket spins around on its central axis. Centripetal force them causes the water to "climb" up the sides of the bucket and stay there for as long as the bucket is spinning, and when the bucket stops spinning, the water falls back to level again.

Now suppose Dexter mounts this bucket on a rigid frame inside a larger, closed container, and fits a camera in the top of the container to record what happens to the water in the bucket. If he spins the container, once again the water will crawl up the sides of the bucket and remain there until the container stops spinning, when it levels out again.

If Dexter then observes the video taken by the camera, it will show the water sitting flat and level in the bucket, then rising up the sides of the bucket for a time, and then falling back level again. Although the container is closed, the action of the water tells Dexter when the container was spinning and when it was not. The subtle problem is: how does the water in the bucket "know" it is spinning? Gravity?

No, gravity pulls straight down on it. The same puzzle arises with simple linear acceleration: motion may be a relative concept, but acceleration is not. How does a mass "know" what it is accelerating relative to? Isaac Newton discussed this puzzle in the course of his work, suggesting that there was some sort of "absolute space" that let the bucket "know" it was spinning, or let a mass "know" it was accelerating.

Newton's concept of absolute space was vague, and not long afterward the Irish philosopher, mathematician, and bishop George Berkeley, following ideas put forth by Leibniz, suggested that the distant stars defined the structure of space, in some way connecting the bucket to the rest of the Universe. After all, if the container and the bucket were the only things in the Universe, could they ever spin? What would they spin relative to? If they could only spin relative to each other, what would the result be? The same thought applied to linear accelerations.

Berkeley was far ahead of his era and nobody paid him much mind at the time. In the 1860s, the Austrian physicist Ernst Mach reinvented the idea. He really didn't add much that Berkeley hadn't said, though Mach did suggest that the same effect would be obtained if the bucket were stationary and the distant stars rotated around it. There was no "absolute space", space being defined only by the masses it contained. This was not an arbitrary distinction, either: if the Universe contained less mass, then liquid in a bucket would not rise as far up the walls when the bucket was spun, and if there were no mass in the Universe, the liquid wouldn't rise at all in effect, spin wouldn't exist.

This little puzzle is known as "Mach's principle", though it should better be known as the "Berkeley-Mach principle". It is a completely maddening thing to consider, and from the point of view of simple practical physics, it's of little or no interest, which is why it's only being discussed as a footnote. The reason it's mentioned here is as a warning against complacency.

As mentioned before, simple physics implies simple models of how the real Universe works. These models can be effective and very useful, and in the best case strip away the clutter of details to focus on the essentials, but the real Universe is substantially more complicated. Anyone who is tempted to become glib about physics needs to understand that on closer investigation, the simple models may not fit the way things actually work in every fine detail. Science, as the saying goes, isn't necessarily an exact science.

Chapter 4

Gravity and Orbital Mechanics

- The most elegant and simple way to observe the action of gravity is to show how it defines the movements of objects in space. In fact, it was studies of the orbits of the planets that led Isaac Newton to devise his formula for gravity in the first place. This chapter gives a short introduction to "orbital mechanics".

ELEMENTARY ORBITAL MECHANICS

- As established in a previous chapter, gravity is the only one of the four forces that operates over long distances. It is an attraction between two masses and its magnitude is given by:

$$\text{gravitational force} = \text{constant} \times \text{mass}_1 \times \text{mass}_2 / \text{distance}^2$$

where the constant has a value of 6.672×10^{-11} in standard metric units and the distance is between the centre of mass of the two objects. The precise value of the very small constant has been determined by painfully exact experiments, in which a pair of masses are balanced in a vacuum on a bar supported by a wire thread, and allowed to pivot towards larger fixed masses. This experiment was originally performed by the English scientist Henry Cavendish in 1771, and has been repeated since then to ever greater levels of precision.

The force of gravity continues to grow weaker as the distance from Earth increases, but it never stops completely. Earth, or any other celestial body in the Universe, is said to be at the bottom of a "gravity well".

- The workings of gravity can clearly considered by visualizing the motion of an artificial satellite in orbit around the Earth. The orbit of such a satellite is defined by the balance of the satellite's orbital velocity against the force of the Earth's gravitational attraction. This force is given by:

gravitational force = constant × mass earth × mass of satellite / distance^2

The mass of the satellite is irrelevant. Double the mass of the satellite and the force of the Earth's attraction on it is doubled, but the amount of force required to get the increased mass to accelerate to a given velocity is doubled as well. Since force is equal to mass times acceleration, then:

gravitational acceleration = constant × mass of earth / distance^2

The acceleration of gravity at the surface of the Earth is 9.81 meters per second squared, and so this formula can be simplified further, to: gravitational acceleration = 9.80665 / (orbital radius / radius of earth)^2

The orbital radius is measured in meters from the centre, not the surface, of the Earth. The radius of the Earth is 6,340,000 meters (6,340 kilometers). This means that the gravitational acceleration of a satellite in an orbit 1,000 kilometers above the surface of the Earth would be:

$$9.81 / ((6{,}378{,}000 + 1{,}000{,}000) / 6{,}378{,}000)^2$$
$$= 7.33 \text{ meters per second squared}$$

In a stable circular orbit, this force is balanced by the centripetal acceleration of the satellite due to its orbital velocity, which is given by: centripetal acceleration = orbital radius × angular velocity^2

Balancing the two expressions gives:

$$9.81 / (\text{orbital radius} / \text{radius of earth})^2$$
$$= \text{orbital radius} \times \text{angular velocity}^2$$

or:

$$\text{orbital radius}^3 = (9.81 \times \text{radius of earth}^2) \times \frac{1}{\text{angular\_velocity}^2}.$$

This equation is a little easier to handle if angular velocity is converted into "orbital period", or the amount of time it takes for a satellite to make one orbit of the Earth, in hours. Radians per second can be converted to revolutions per second by dividing by 2π, and then revolutions per second can be converted to revolutions per hour by multiplying by $60 \times 60 = 3{,}600$. Orbital period is the inverse of revolutions per hour, and so the conversion is:

$$\text{angular velocity} = \frac{2PI}{3{,}600 \times \text{orbital\_period}}$$

Plugging this into the equation gives: orbital radius^3

$$= (9.81 \times \text{radius of earth}^2) \times \frac{1}{(2PI / 3{,}600 \times \text{orbital\_period})^2}$$

$$= \text{orbital period}^2 \times \frac{3{,}600^2 \times 9.81 \times \text{earth\_radius}^2}{2PI^2}$$

Giving the orbital radius in meters is also inconvenient, but that can be adjusted to give it in kilometers by dividing through by $1{,}000^3$:

$$\text{orbital radius}^3 = \frac{3{,}600^2 \times 9.81 \times \text{earth\_radius}^2}{1{,}000^3 \times 2PI^2} \times \text{orbital period}^2$$

This leads to a very simple relationship:

$$\text{orbital radius}^3 = \text{constant} \times \text{orbital period}^2$$

This is known in general as "Kepler's Third Law". Kepler's three laws are a description of planetary orbits devised by the German astronomer Johannes Kepler from observations of such orbits. They strongly influenced the development of Newton's law of universal gravitation. The other two laws are discussed in later sections of this chapter.

Kepler's Third Law is true for any planetary orbit. In the case of an Earth satellite, the value of the constant is 130,958.995396 (using the more precise value of 9.80665 for the acceleration of gravity instead of 9.81), and the law can now be rearranged to give two simple formulas:

$$\text{orbital radius} = 5{,}078 \times \text{orbital period}^{(2/3)}$$

$$\text{orbital period} = (\text{orbital radius} / 5{,}078)^{1.5}$$

with the orbital period in hours and the orbital radius in kilometers.

SATELLITE ORBITS

- Kepler's Third Law can be used to calculate simple satellite orbits. For example, for a satellite to have a period of 2 hours, it will have an orbital radius of:

$5,078 \cdot 2^{(2/3)} = 8,060$ kilometers

This corresponds to an altitude of 8,030 – 6,340 = 1,690 kilometers. Similarly, at an altitude of 19,000 kilometers, or three Earth radii, the satellite has a period of:

$((6,378 + 19,000) / 5,078)^{1.5} = 11.2$ hours

The fact that a higher orbit is a slower orbit, at least in terms of angular velocity and period, was confusing to astronauts performing the first space rendezvous operations in the mid-1960s. They would be trying to catch up with a target spacecraft, but simply firing their rockets directly ahead would push them into a higher orbit, where they would find themselves falling *behind* the target. They were all originally trained as pilots; it was nothing like flying an aircraft and was completely contrary to their instincts. They adjusted quickly, learning to direct their boost in an angle below the line to the target to modify their orbital direction as well as their speed.

An orbit with a period of 24 hours has an interesting property, in that this is the same as the rotation period of the Earth, and so the satellite "hangs" over one spot on Earth. There is a minor trick in calculating this because the motion of the Earth around the Sun makes the day about 4 minutes longer than it would be otherwise. The "sidereal" day relative to the stars is 23.93 hours. That means the geostationary altitude is:

$5,078 \times 23.93^{(2/3)} = 42,172$ kilometers

or an altitude of about 35,800 kilometers. Of course, since satellites orbit around the centre of mass of the Earth, such a "geosynchronous" or "geostationary" orbit has to be over the equator, or the satellite will weave north and south of the equator during its 24-hour orbit. Geostationary orbits are very handy for some types of satellites, such as wide-area weather observation and, particularly, communications satellites.

The orbital "slots" where a satellite may be placed in geostationary orbit are regulated and are traded as a commercial commodity by satellite companies. A satellite TV user has to focus his or her home dish antenna on a satellite in a particular orbital slot and then lock the dish in place.

By the way, the Earth's axis of rotation is at an angle to its orbit around the Sun, and so the plane of the orbits of geostationary satellites around the equator are at an angle to the plane of the Earth's orbit as well. This means that the Sun's pull tends to tug the satellites out of position, and so they are fitted with small "station keeping" thrusters to keep them in their position. These thrusters do not need to be very powerful, since only a slight nudge is needed. Once the fuel for the thrusters is gone, the satellite will gradually drift out of its geostationary orbit and become useless.

- Not all satellites are put into geostationary orbit, of course. 35,800 kilometers is a long way from Earth. Communications satellites that have to pick up signals from small ground transmitters that don't have the power to reach geostationary orbit are implemented as "constellations" of small satellites in fast low Earth orbits (LEO), with enough of the satellites in orbit to make sure one that when one goes down over the far horizon, another has come up over the near horizon to make sure communications remain constant.

Satellites that observe the Earth in detail for military, commercial, or scientific purposes are also usually in LEO, placed into orbits over the poles, so that as they spin around the Earth once every hour or so, the Earth rotates underneath them. In 24 hours, such a satellite will be able to scan the entire Earth.

In reality, while a strictly polar orbit might be used for satellites that scan the Earth with radar, those that take pictures are not usually put into an orbit that takes them precisely over the poles. They are put into a slightly offset orbit at a specific altitude, with the interesting property of placing them over a particular location on the equator at the very same time every day in a particular time zone.

This is known as a "Sun synchronous" orbit, and it is donc to ensure that the lighting conditions are as consistent as possible between observational passes. Changing shadow positions and other lighting conditions makes comparison of images from successive passes of a satellite difficult. If the shadows are the same on each orbit, then any changes are much more noticeable; if shadows suddenly appear in a frame taken by a military reconnaissance satellite and weren't in the previous frame, it might mean that equipment has been moved into that location since the last pass over the target. For example, Argentina's SAC-C scientific Earth observation satellite, launched late in 2000, was put into orbit at an altitude of 702 kilometers and an inclination of 98.2 degrees from the equator. The inclination is described as 98.2 degrees, instead of 81.8 degrees, to emphasize that the satellite has a "retrograde" orbit, that is, in the reverse direction to the Earth's rotation, as opposed to "prograde" or in the same direction of the Earth's rotation. SAC-C passed over the equator at 10:15 AM local time during each orbit.

- The orbits described so far are circular, but there's also no reason that they can't be elliptical, dropping down close to the Earth at one part of its orbit, and then rising very high over the Earth at another part of its orbit.

Such an elliptical orbit demonstrates an interplay between kinetic and potential energy. At the lowest part of its orbit, the "perigee", the satellite is moving very fast. It has maximum kinetic energy and minimum potential energy.

As it circles around the Earth and arcs to greater heights, it loses kinetic energy to gravitational potential energy and slows, eventually reaching a minium velocity at the very top of the ellipse, the "apogee". It now has minimum kinetic energy and maximum potential energy, in much the same way as a ball thrown upward reaches the top of its arc, floating for an instant before reversing its direction. Once the object in orbit reaches its apogee, it then falls back down along its orbit, picking up speed until it reaches the perigee again.

Incidentally, the terms "perigee" and "apogee" specifically refer to an orbit around the Earth. For an orbit around the Sun, the terms "perihelion" and "aphelion" are used; for an orbit around a star, the terms "periastron" and "apoastron" are used; and if the central body isn't specified, the terms "periapsis" and "apoapsis" are used.

In any case, the behaviour of an object in an elliptical orbit leads to a neat relationship: if a radius is drawn from the spacecraft to the centre of the Earth, that radius will sweep out equal areas in equal intervals of time. This is "Kepler's Second Law".

If the satellite is near its apogee, that radius is very long, and since it is moving slowly in a particular interval of time the area it sweeps out is a long, narrow slice. If the satellite is near its perigee, it is moving fast, and the area it sweeps out in the same amount of time is a short, wide slice with exactly the same area.

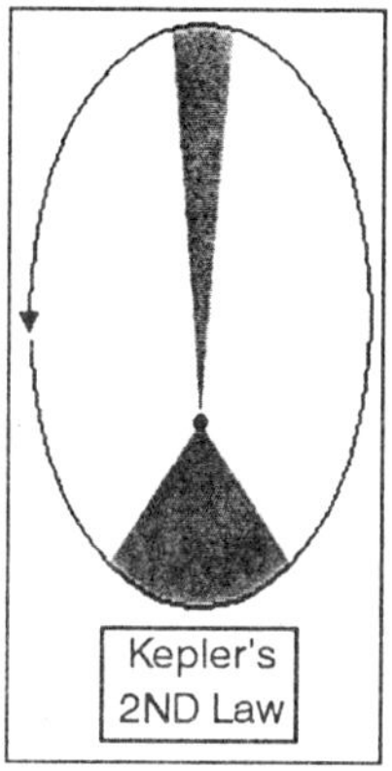

Elliptical orbits were particularly useful to the Soviet Union. The USSR was at high latitudes, and geosynchronous communications satellites tended to be too close to the horizon to be useful in the northern regions of the country. As a result, the Soviets launched communications satellites with the name of "Molniya (Lightning)" into elliptical orbits that took them high over the USSR, arcing overhead for long periods of time, and then falling back down around the other side of the Earth to make whip around quickly for another gradually slowing arc up over the Motherland.

This orbit became known as a "Molniya orbit". The Americans employed Molniya orbits for spy satellites used to eavesdrop on Soviet communications. Although the modern Russian state seems to be more interested in geostationary communications satellites, some American eavesdropping satellites are still launched into Molniya orbits.

- Although everyone knows an ellipse is a "less than perfect" circle, it is formally defined as the points on a closed curve where the sum of the distances from two "focal points" within the curve to one of the points on the curve are a constant.

A more intuitive way to see this is to imagine two pins stuck into a piece of cardboard to act as focal points, with a loop of thread laid around the threads. A pen is used to stretch the loop taut and then trace a curve all the way around the two focal points. Obviously, if both focal points are in the same place, the pen will draw out a circle, but the farther apart the pins are placed, the more elliptical the curve becomes.

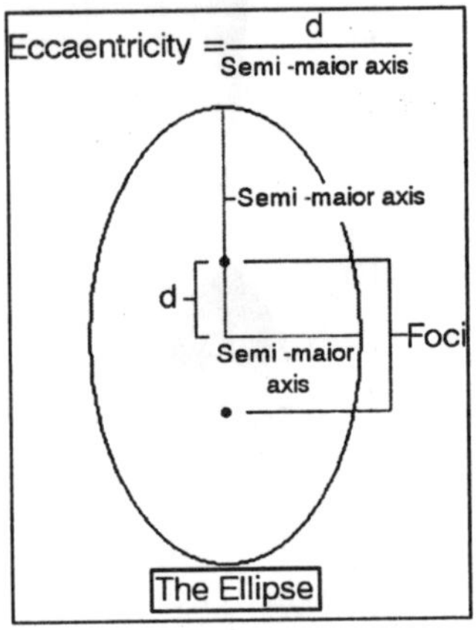

The longest distance from the edge of the ellipse to its centre is called the "semi-major axis", while the shortest distance is unsurprisingly called the "semi-minor axis".

The "eccentricity" is given as the ratio of the distance from the centre to a focal point versus the length of the semi-major axis. This has a value of 0 for a circle, with the focal point at the centre of the curve, and approaches a value of 1 as the curve becomes more elliptical.

The semi-major axis is often used as the "average value" of an orbital radius, which makes sense since it's the average of the maximum and minimum distance to the central planet or other body.

Suppose that a hypothetical moon of some equally hypothetical large planet has an apoapsis of 7.5 million kilometers and a periapsis of 6 million kilometers. This is a fairly big orbit and so the diameter of the planet is disregarded. This gives the semi-major axis as:

$$(7.5 + 6) / 2 = 6.75 \text{ million kilometers}$$

The distance from the centre of the orbit to the focal point where the planet resides is:

7.5 – 6.75= 0.75 million kilometers = 750,000 kilometers

This gives the eccentricity of the orbit as:

$$0.75 / 6.75 = 0.11$$

Similarly, suppose another moon of the same planet is said to have an orbit with an average value of 9 million kilometers and an eccentricity of 0.15. Interpreting the average value as meaning the semi-major axis, then the distance from the centre to a focal point is:

$$9 \times 0.15 = 1.35 \text{ million kilometers}$$

This gives the apoapsis as:

$$9 + 1.35 = 10.35 \text{ million kilometers}$$

and the periapsis as:

$$9 - 1.35 = 7.65 \text{ million kilometers}$$

Of course, the orbital inclination of the moons relative to the planet's equator also has to be specified to define their orbits. Incidentally, moons that were created along with a planet usually have prograde, nearly circular orbits in the plane of the planet's equator, while moons that were captured later usually have strongly elliptical orbits that can be prograde or retrograde, at almost any inclination to the planet's equator.

How the moons were captured later is a bit of a puzzle, since if they fell in towards a planet from deep space and didn't hit it, by energy conservation they would simply swing back out to deep space from whence they came and not go into orbit around the planet.

It is believed that planets with captured moons were surrounded by an envelope of gas and dust in their early days that imposed enough drag on the moons to capture them. Also incidentally, there is a simple relationship between the semi-major axis, semi-minor axis, and distance from the centre to a focal point (FP): semi-major axis^2 = semi-minor axis^2 + distance to FP^2

- The best place to launch an Earth satellite into a geostationary orbit is from a launch site near the equator, such as the European Space Agency's space centre in Kourou, French Guiana, on the northern coast of the South American landmass; or the "Odyssey" floating platform used by the Sea Launch commercial rocket group to boost satellites into orbit from the equatorial Pacific. An equatorial launch not only minimizes the amount of fuel needed to bring the satellite into equatorial orbit, but the high linear velocity of the Earth's rotation at the equator gives a launcher an extra "boost" into space.

The Earth also has a slight "bulge" at the equator, with the equatorial diameter being about 42 kilometers more than the polar diameter, though this is a minor consideration in space launches.

However, one implication of the irregular shape of our planet is that the standard value of the Earth's gravitational acceleration, 9.81 meters per second squared, varies a bit depending on location from about 9.83 at the North Pole to about 9.78 on the Equator. On top of Mount Everest, it actually drops down to almost 9.76.

After launch, a geostationary satellite is placed into a highly elliptical temporary "geostationary transfer orbit" with the apogee at 35,800 kilometers. Once the satellite reaches that altitude, it fires a rocket engine to "circularize" its orbit and not fall down to its perigee again.

The concept of transfer orbits was described by the German space pioneer Walter Hohmann in 1925, and so this is an example of a "Hohmann transfer orbit". An equatorial launch site is the worst choice for a polar orbit, since the launcher's upper stage must expend fuel to turn the spacecraft's orbit towards the poles.

The launcher will have to expend additional fuel to provide the change in velocity, or "delta vee", to get the payload into its proper orbit, and the weight of the additional fuel means a smaller payload weight. A high-latitude space centre is better for launching such spacecraft. The primary American space launch centers are the Kennedy Space Centre in Florida, and Vandenberg Air Force Base in California.

Since Kennedy is substantially closer to the equator than Vandenberg, Kennedy is used to launch spacecraft into geostationary orbit, while Vandenberg is used to launch spacecraft into polar orbit.

Similarly, the two main Russian launch centers are Baikonur (Tyuratam) in Kazakhstan in Central Asia and Plesetsk in northern Russia, northeast of Saint Petersburg; Baikonur is used to launch payloads into geostationary orbit, while Plesetsk is used to launch payloads into polar orbit. It is not impossible to launch a geostationary payload from high latitudes, nor is it impossible to launch a polar payload from low latitudes, but it is inefficient.

- As this explanation of satellite orbits shows, they are carefully planned and well defined. This brings up as a counterexample the memory of a thriller novel a spy satellite moving around sideways relative to its current orbit and then stopping over a particular target to take pictures. Many low-orbit spy satellites can in fact maneuver, but only to adjust their orbital paths to eventually take them over specific targets.

Obviously, the author was under the popular misconception that the force of gravity suddenly stops working when a spacecraft gets above the atmosphere. The reality is that the spacecraft is falling all the time, it's just that its forward velocity is great enough to allow it to fall over the horizon before it hits ground. If it stopped in orbit, it would drop like a brick. Considering that I don't remember anything else about the novel, at least the author gave me a bit of unintentional humour to remember him by.

INTERPLANETARY SPACECRAFT TRANSFER ORBITS

- Although the Earth's gravity well goes on forever in principle, that doesn't mean it takes infinite energy for a spacecraft to leave the Earth: one of the elementary but subtle truths of mathematics is that many curves that go on forever actually may be summed up to a specific value. If a spacecraft is launched at a high enough velocity, called an "escape velocity", it will have enough energy to ensure that it will never be pulled back into the Earth's gravity well again. The escape velocity for Earth is 11.2 kilometers per second.

To send a space probe to another planet, the probe has to be blasted out of the Earth's gravity well at more than the escape velocity. However, it isn't enough to just point the launch rocket at the planet and blast away. Even the largest booster rocket can only burn for times measured in a total of minutes, while the planets are so far away that a probe will take months or years to reach them. This means that the probe has to be launched on a trajectory that "leads" the motion of the target planet. The probe is launched out of the Earth's gravity well into an orbit around the Sun that arcs between the Earth's orbit and the target planet's orbit. This trajectory is called a "Hohmann interplanetary transfer orbit".

To reach Mars, the probe must be given a delta-vee that increases its velocity relative to the Earth's orbit around the Sun so that its transfer orbit will reach outward towards Mars. Counterintuitively, to reach Venus, the probe must be given a delta-vee that *reduces* its velocity relative to the Earth's orbit so that the transfer orbit will fall inward towards Venus. In other words, the launch vehicle actually *slows down* the spacecraft, at least relative to its orbit around the Sun.

Software is used to calculate transfer orbits. Such programs are by no means trivial to write. The problem is that there are many bodies in the Solar System, and a spaceflight engineer has to at the very minimum consider the interactions of the Earth, the target planet, and the Sun to properly calculate a transfer orbit.

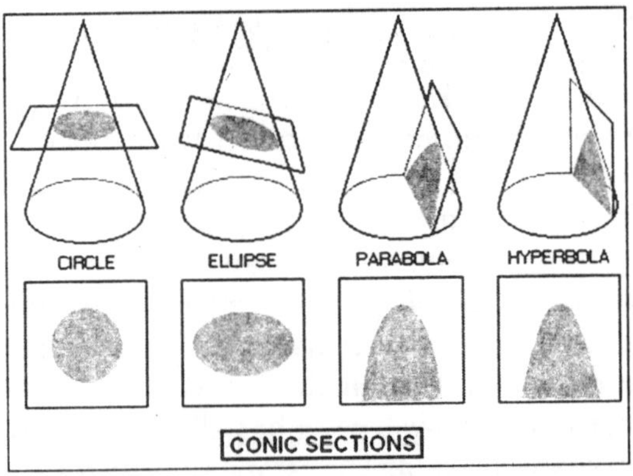

The gravitational interactions of three or more bodies are so complicated that they are computationally overwhelming, and to make matters worse such systems are "nonlinear": Any slight change in initial conditions can quickly lead to wild variations in trajectory. This is known as the "N-body problem."

In practice, transfer orbits are calculated using simplified methods. For example, calculating the interactions of two bodies is an exercise in simple physics. The trajectories are defined by "Kepler's First Law", which states that motions of a planet or a spacecraft around the Sun or other larger mass are always in the form of a "conic section".

A conic section is the geometric figures that can be obtained by cutting through the side of a hollow cone and tracing out the edge of the cut. A circle (eccentricity = 0) is obtained by cutting through the cone at an angle perpendicular to the axis of the cone.

Move the angle away from the perpendicular, the section becomes an ellipse (eccentricity < 1) that becomes more elongated as the angle increases. Once the angle becomes exactly that of the slope of the cone, the ellipse becomes an open figure known as a "parabola" (eccentricity = 1). For steeper angles up to the point where the cut is parallel to the axis, the open figure widens into a series of "hyperbolas" (eccentricity > 1).

The movements of celestial bodies are never perfect circles or parabolas, these being "perfect" figures; they are ellipses when they are in repeating orbits around a body, and hyperbolas when falling in from deep space on a nonrepeating orbit. Orbital planning software calculates one set of conics to define the spacecraft's trajectory at launch, calculates a second set of conics for the spacecraft's trajectory at its target, and tries to find one conic in each set that match up.

- When the probe arrives at the target planet, it will generally have so much velocity that it will swing by the planet and continue on it orbit around the Sun. Traditionally, the probe has to fire its rocket engine to reduce its velocity enough to allow it to be captured by the planet's gravity.

  This requires a fair amount of fuel, and a more economical approach known as "aerocapture" has been considered. In this scheme, the probe is fitted with a heatshield and actually skims through the upper atmosphere of the planet, presuming it has one, to literally "burn off" the excess velocity using friction.

This is unsurprisingly a very precise operation. Although "aerobraking" was used to circularize the orbit of the American

Mars Global Surveyor (MGS) probe launched in 1996, MGS used its rocket engine to allow it to be captured by Mars' gravity in the first place. So far no spacecraft has successfully performed "aerocapture", or skimming a planet's atmosphere to reduce velocity enough to be captured by the planet's gravity.

- Spaceflight enthusiasts have used transfer orbit calculations to come up with an ingenious idea known as the "cycler". This is a space station that is in a permanent transfer orbit between, say, Earth and Mars. The cycler would pass close by Earth and later pass close by Mars on a regular schedule. Astronauts would be able to hitch a ride on the cycler station to go to Mars, and later pick up the cycler again to return to Earth. The astronaut's trip times for an Earth-Mars cycler would be several years, but some schemes fit the cycler with a high-efficiency rocket system to adjust the trajectory, permitting trip times as low as half a year.

GRAVITY ASSIST TRAJECTORIES

- A simple Hohmann interplanetary transfer orbit works well enough to send a probe to Mars or Venus, but sending a probe to the distant outer planets using such a trajectory would take far too much time. Space launch planners have developed an interesting and subtle way to "cheat" and greatly reduce the time and expense of such missions by sending the probe on a "slingshot" trajectory around a planet to obtain a "gravity assist".

This is a subtle idea. If a probe is sent on a flyby around a planet, it approaches the planet from one direction, arcs around it, and departs in another direction. As viewed from the planet, the probe will not leave the planet with any greater velocity than it arrived with. The probe has the same energy before and after the encounter, even if it speeds up during the flyby.

The trick is that the planet being used for the gravity assist is a moving object. If the probe swings behind the planet in its orbit, some of the velocity of the planet is transferred to the probe while the planet slows down slightly. However, since the velocity transfer between the two objects is inversely proportional to their mass, the probe's increase in velocity is substantial, while the planet's decrease in velocity is literally unmeasurable. A probe can be swung around in front of a planet in its orbit to decrease velocity. A gravity assist can be regarded as a "soft collision", with the spacecraft "bouncing" off a planet and obtaining a delta-vee thereby.

- The concept of a gravity assist trajectory was pioneered in the 1960s by engineers at the US National Aeronautics and Space Administration's Jet Propulsion. Laboratory (NASA JPL) in California. By a stroke of good luck, their work on gravity slingshots led them to realized that beginning in the late 1970s, all four outer gas giant planets would be strung out in a perfectly staggered formation beginning in the late 1970s, a configuration that only occurred once every 175 years. The new gravity assist technique could be used to slingshot a space probe from one planet to the next.

This "Grand Tour" trajectory was implemented on the Voyager 2 space mission, which was launched in 1977. The Voyager 2 probe performed a flyby of Jupiter in 1979, slinging it on to a flyby of Saturn in 1981, a flyby of Uranus in 1986, and a flyby of Neptune in 1989. Voyager 2's scientific payback was unexceeded by any other single space mission. Gravity assist trajectories have been used on many other deep-space missions. For example, the Galileo Jupiter orbiter was faced with being drastically cut back in the mid-1980s because the powerful booster rocket originally designed for it was cancelled, and a much smaller rocket had to be used instead.

Fortunately, the mission was saved by coming up with a "Venus-Earth-Earth Gravity Assist (VEEGA)" trajectory in which the probe swung past Venus once and Earth twice to obtain a gravity assist on each pass. After one Galileo Earth flyby, JPL officials were startled to see protesters waving signs complaining about the way Galileo was robbing Earth of its orbital momentum.

It quickly turned out the protesters were really just a gang of jokers who were being silly on purpose, partly to mock more earnest protesters who had previously objected to the launch of Galileo because its atomic-fueled electrical power units.

GRAVITY AND PLANETS/TIDES

- The orbital scenarios discussed so far have involved a great big planet and, in comparison, a very small spacecraft. The basic concepts aren't different when two planet-sized bodies are involved, but some additional rules come into play.

The first additional rule is that one body doesn't orbit around the other: they both orbit around the centre of mass of the system. A spacecraft in orbit around the Earth has so little mass that its effect on the Earth is negligible, and so the spacecraft's orbit for all practical purposes goes around the centre of the Earth. However, if two planet-sized bodies of the same size orbit each other, then they both orbit around a common point halfway between each other. This point is sometimes referred to as the "barycenter".

If they are less equal in size, the barycenter is their centre of gravity. A good example of this is the distant world Pluto and its large moon Charon, Charon has a mass that is 15 per cent that of Pluto's, and the distance between the two is about 20,000 kilometers. The balance point P, as measured from the centre of Pluto, can be determined with the simple equivalence:

mass of pluto × P = (20,000 – P)×mass of charon

mass of pluto×P + mass of charon × P = 20,000 × mass of charon

P×(mass of pluto + mass of charon) = 20,000 × mass of charon

P = 20,000 × mass of charon / (mass of pluto + mass of charon)

= 20,000×0.15/(1+ 0.15) = 2,610 kilometers

The second additional rule is that the two bodies have "tidal" effects on each other, due to the differential effect of gravity across the diameter of a planet. The Moon and Earth have a strong tidal interaction on each other, manifested on Earth as the rise and fall of ocean coastal waters during the passage of a day.

Although the sum of the gravitational force between the Earth and the Moon acts on a line through their centers, of course the gravitational force of each world acts on every part of the other. Since gravitational force decreases with the square of the distance, the Moon's force on the near face of the Earth is weaker than its force on the far face. This difference in force between the two faces is equivalent to a force between them that attempts to stretch the Earth into an ovoid, with its long axis pointed towards the Moon. Since the Earth is pretty solid, the tidal effect on the land masses is slight, but the tidal strain is enough to cause the oceans to shift around on a regular cycle.

While the usual tidal change is a meter or two, some coastal estuaries funnel the tides and increase their height to up to 15 meters, and hydropower dams have been built in a few such locations. When the tide goes high, the dam opens to let the seawater into the estuary, spinning a power turbine on the way in. The dam is then closed, and when the tide drops, the dam opens to let the water back out into the ocean, spinning the turbine on the way back out again. The Sun also contributes to tides, but the although the Sun is much more massive than the Moon and has a much greater gravitational force on the Earth, the Sun is much farther away. An analysis of tides as a differential gravitational force shows they fall off with the *cube* of distance between two masses, not the square as does gravity itself, so the Sun's tidal influence falls off much faster than its gravitational influence.

Higher tides do occur when the Sun and Moon are aligned on opposite sides of the Earth, causing their tidal effects to work together. These are known as "spring tides", meaning they "spring up" above normal, not because they happen in the spring they can happen in any season. When the Sun and Moon are at right angles relative to the Earth, their tidal effects work against each other, and the result are low or "neap" tides. Tides also apply to the Earth's atmosphere, with atmospheric pressures varying (on the average) from minima to maxima in step with high and low tides. Oddly, the Sun has a greater effect on atmospheric tides than the Moon. This is because, as explained in a later chapter, warm gases tend to expand, and solar heating drives that expansion, magnifying the solar tidal effect.

- One of the interesting implications of the tidal forces on the Earth is that they cause precession of the Earth's spin axis. The Earth's spin axis is tilted at an angle of 23 degrees to the ecliptic, the plane of the planet's orbit around the

Sun, while the Moon's orbit is in the plane of the ecliptic. The tidal forces of the Moon and Sun tug the spin axis of the Earth back towards the horizontal, but since the Earth is spinning this causes precession instead, with a period of about 26,000 years. At the present time, the Earth's spin axis points towards Polaris, the pole star, but 13,000 years from now it will point to a location near the bright star Vega.

The Moon's Consta Face (GVG/PD)

The effect of the Earth's tidal force on the Moon is much more dramatic and is easily seen by the fact that the Moon only presents one face to the Earth at all times. Over the eons, the Earth's tidal effect slowed the Moon's rotation until it became the same as the period of the Moon's orbit around the Earth. The Moon is said to be "tidally locked" to the Earth. Incidentally, the Moon's tidal forces are very gradually slowing down the Earth's rotation as well. 900 million years ago, the Earth's day was only 18 hours long. The Moon was 7.5 per cent closer to the Earth, and the lunar month was 23.4 days.

The distance between the Earth and the Moon can now be precisely measured by bouncing a laser beam off laser reflectors left on the Moon by the Apollo manned lunar expeditions. These measurements show that tidal stresses set up by the Earth on the Moon are accelerating it, and it is moving farther away from the Earth at a rate of 3.8 centimeters per year.

Chapter 5

Concepts in Orbital Mechanics

GRAVITY IN THE SOLAR SYSTEM / LIBRATION POINTS / STELLAR MASSES

- The previous chapter discussed how the Earth and Moon are tidally locked, with the Moon always keeping its same face to Earth. Most of the major moons in the Solar System are also tidally locked to their planets. In the Pluto-Charon system, both planets are tidally locked to each other, so Charon hangs overhead in Pluto's sky in the same place at all times.

  The energy of rotation of the two worlds was dissipated in internal heating due to tidal stretching effects. This tidal heating appears to be a significant process in the internal heating of Jupiter's volcanic moon Io, which is relatively close to the giant planet and so is subjected to strong tidal forces. The surface of Io is covered with huge sulfurous volcanic flows.

It was once thought that the planet Mercury was tidally locked to the Sun, so that it kept one face eternally faced to the Sun and the other in darkness, one side burning and the other side freezing. This was found to be not exactly the case in the 1950s. It is in a "resonance" where it turns three times for every two orbits around the Sun, in what is known as a "metastable" state. If Mercury kept one face always to the Sun, it would be in a minimum energy state, but to get out of its 2:3 resonance state it would have receive some sort of increase in rotational energy, possibly through a collision with another large body.

Until something like this happens, Mercury's rotation will

remain as it is, something like a stone that been caught on a ledge after falling halfway down a cliff instead of dropping all the way to the bottom.

- Due to the N-body problem, the gravitational interactions between all the planets in the Solar System are complicated and difficult to predict over the long run. At a coarse scale, the gravitational interactions of the Solar System are dominated by the Sun, which is over 330,000 times more massive than the Earth; and Jupiter, which is about 320 times more massive than the Earth. The second biggest planet,
  Saturn, is only about 95 times the mass of the Earth, and all the other planets are much smaller than Saturn. Jupiter has well more than twice the mass of all the other worlds in the Solar System combined.

Jupiter's gravitational influence led to the creation of the asteroid belt, since it prevented the formation of any single large body at that location. Interestingly, there are empty "bands" in the asteroid belt that correspond to asteroid orbits that would have a some sort of synchronization or resonance with the orbit of Jupiter. Any asteroid in such an orbit is quickly pulled out of it by periodic gravitational tugs from Jupiter.

Another consequence of Jupiter's gravitational influence is that objects straying into the two locations in its orbit that are equally distant from the Jupiter and the Sun tend to stay there. These orbital positions are known as "libration points" note the spelling, it's easy to misspell as "libation" or sometimes "Lagrangian points" after the French mathematician Louis Lagrange.

Lagrange performed an analysis of the gravitational interactions of two bodies in orbit around each other to identify positions where objects would remain stationary relative to the bodies. He actually discovered five such stable positions, now known as "L1" through "L5", roughly laid out as follows relative to two masses of the same size:

L1, L2, and L3 are in unstable equilibrium, only stable in the sense that a ball balanced delicately on a needle point is stable. L1 is located between the two bodies, and an object placed there will fall towards one or the other if disturbed. Objects at L2 and L3 are at a precise orbital radius around the two bodies where they keep pace with the orbits of the bodies. At a lower radius, the objects will have a faster orbit and move ahead, while at a higher radius, they will have a slower orbit and fall behind.

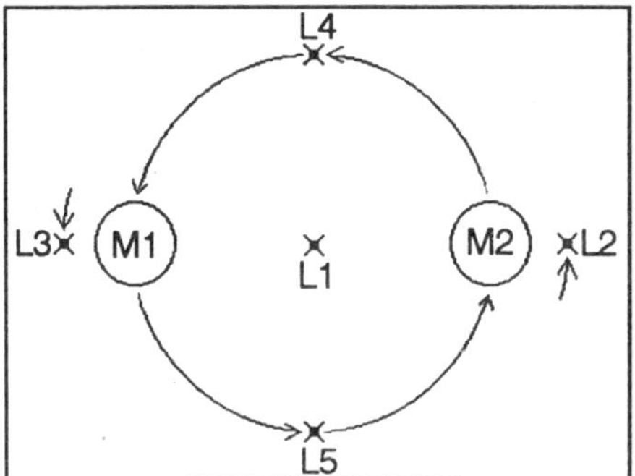

Libration Points

In contrast, the equidistant positions, L4 and L5, are in stable equilibrium, just as much a ball caught in a depression is in stable equilibrium. If any object caught at these locations is disturbed and begins to move out of them, the forces of the two bodies tend to nudge it back into place. The L4 and L5 positions in Jupiter's orbit have trapped a number of asteroids, referred to as "Trojan asteroids".

- One effect of the interactions between the planets is to rule out the existence of a "Counter Earth", a plot device found in old science-fiction stories. The idea is that there is another planet on the other side of the Sun exactly opposite to Earth that remains eternally unseen.

  This would seem implausible in the first place, since it's obvious there are no other "counter planets" such as a "Counter Mars" or "Counter Jupiter".

  The Earth is on the opposite side of the Sun relative to every other planet at some time or other, and no counter-worlds for Mars or Jupiter have been seen. Furthermore, a Counter Earth would be a bright object, and no deep space probe has ever spotted such a thing with its telescopic cameras or other sensors.

In reality, a Counter Earth is completely impossible, since even if such a world were magically created right now, the unbalanced interactions of all the other planets on the Earth and Counter Earth would disrupt the neat balance, possibly leading to a cataclysmic collision between the two planets.

However, the Earth does actually share its orbit, literally more or less, with a handful of small asteroids. The discovery of the first of these "Earth coorbital" asteroids, designated "3753 Cruithne", a Celtic name pronounced "crou-EEN-ya", was announced in 1997. 3753 Cruithne's orbit is known as a "horseshoe" orbit, which describes the path its orbit traces out if the Earth is regarded as stationary.

To visualize the simplest form of a horseshoe orbit, imagine an asteroid that is orbiting slightly inside the orbit of the Earth. As it is in a tighter orbit, it orbiting faster than the Earth, and gradually catches up to it.

This is where things get interesting. Once the asteroid gets close to the Earth, the Earth's gravity nudges it into an orbit outside that of the Earth's. The asteroid is now moving more slowly than the Earth and falls behind. Eventually, the Earth will catch up to the asteroid in this exterior orbit and shunt it back to the original interior orbit.

Since it is moving faster than the Earth in the interior orbit, it moves away from the Earth, to eventually come around behind it again... and repeat the cycle once more. From the point of view of the Earth, the asteroid's path forms a "horseshoe", continuously approaching the Earth and being sent back on its path again.

As the Earth and the asteroid have similar orbital velocities, it takes centuries for the asteroid to complete a single horseshoe orbit. 3573 Cruithne's orbit actually has a slightly different inclination and orbital eccentricity to that of the Earth, and so from the point of view of the Earth, the asteroid makes a series of many kidney-bean-shaped spirals along its horseshoe path. It performs one such spiral every year, with each spiral taking it one step towards or away from the Earth, and the cycle takes a total of 385 years.

Its orbit actually can overlap that of the Earth, at least as seen from above the plane of the Earth's orbit, but due to the different orbital inclinations of the two worlds it cannot come even as close to the Earth as the Moon.

- The gas giant outer planets have elaborate moon and ring systems that provide interesting examples in orbital mechanics. For one example, Tethys, a moon of Saturn a bit over 1,000 kilometers in diameter, has two small coorbital moons, Telesto and Calypso, both a few tens of kilometers in diameter, which are at the stable libration points, equidistant from Tethys and Saturn.

Saturn also has a set of coorbital moons, Janus and Epimetheus, both with diameters of a few hundred kilometers, that have a relationship very similar to that of the Earth with its coorbital asteroids. The orbits of Janus and Epimetheus are not exactly the same, being about 50 kilometers apart. Of course, the moon in the lower orbit has a faster angular velocity than the moon in the higher orbit, and so the two moons draw together. Once about every four Earth years, they pass each other and trade orbits, with the moon in the lower orbit going into the higher orbit and the moon in the higher orbit going into the lower orbit. Astronomers were a bit confused for a few years until they finally figured out what was going on.

Resonances between the orbits of moons in these systems are known as well. For example, Saturn has two moons named Dione and Enceladus that have a 1:2 resonant relationship, with the outer moon Dione performing one orbit of Saturn for every two orbits of the inner moon Enceladus.

- A knowledge of orbital mechanics is very useful for the studies of distant star systems.

  The stars look like points of light in even the largest conventional telescope, and so astronomers have to rely on sometimes very subtle indirect means to determine their distance, size, and true brightness.

This task is simplified if astronomers find a "double star" system, in which two stars orbit around each other. It is possible to precisely time how long the orbit takes through observation. It is also possible to determine how fast the stars are moving in their mutual orbit through changes in the colour of their light emission, or "Doppler shift", discussed later.

If the velocity of the stars in their mutual orbit is known and the period of the orbit is known, then the length of the circumference of the orbit can be determined by multiplying the velocity times the period, giving the length of the circumference of the orbit. Comparing the actual size of the orbit against its size as seen from Earth can give a good estimate of the distance to the double-star system.

Furthermore, knowing the size of the orbit and its period gives the combined mass of the two stars, using Kepler's Third Law:

$$\text{orbital period}^2 = \text{constant} \times \text{orbital radius}^3$$

The constant in this expression is related to the force of gravity between the two stars and so two masses involved, and so the expression can be algebraically manipulated to give:

$$\text{mass of system} = \text{constant} \times (\text{radius}^3) / (\text{period}^2)$$

The constant can be simplified by specifying the orbital radius in terms of the Earth's orbit, or "astronomical unit (AU)", the orbital period in terms the Earth's orbital period, or years, and the system mass in terms of multiples of the mass of the Sun. For the case of the Earth-Sun system, this gives the system mass, the radius, and the period all as 1, and so the constant has to be 1 to get the right result. Of course, for the Earth-Sun system the mass of the Earth is negligible compared to that of the Sun.

Suppose a distant star system is discovered featuring two stars with an orbital radius of 2.3 AU and an orbital period of 1.2 years. This gives the combined mass of the two stars as:

$$(2.3^3) / (1.2^2) = 8.45 \text{ solar masses}$$

Locating the centre of the mutual orbit will allow the proper masses to be applied to each of the star: if the centre of the orbit is halfway between the two stars, then each weighs 4.225 solar masses.

LIBRATION POINTS, HALO ORBITS, AND MANIFOLDS

- Libration points are very useful in spaceflight planning. The L4 and L5 locations relative to the Earth and Moon have been promoted as ideal locations for space colonies. Similarly, the Sun-Earth L1 position, a hundredth of the way from the Earth to the Sun, is an ideal location for satellites that observe the Sun; and the Sun-Earth L2 position, 1.5 million kilometers beyond the Earth's orbit around the Sun, is an ideal location for satellites used to perform deep space astronomy.

Several solar observatories have been launched into the Sun-Earth L_1 position. At the L_1 position, these spacecraft have an excellent view of the Sun that is never blocked by the Earth, and are within easy communications range of the Earth. In August 2001, NASA launched a spacecraft named "Genesis" that was placed there between Earth and Sun to capture particles emitted by the Sun for return to Earth.

In the summer of 2001, NASA launched the "Microwave Anisotropy Probe (MAP)", which was parked in the Sun-Earth L2 position, far beyond the Moon's orbit away from the Sun. There, the spacecraft began mapping low-level energy emissions from the distant Universe, deploying a shield to block out emissions from the Sun and Earth to its back that might disrupt the measurements. Other astronomical spacecraft are now being designed to take up station at the L2 position.

- Of course, sending several spacecraft Sun-Earth L1 or L2 position would lead to clutter, and the L1 position has the added difficulty that any spacecraft sitting there is in the full glare of the Sun as seen from Earth. The Sun generates radio noise as well as light, and this creates interference in communications with the spacecraft. The solution to both these problems is what is known as a "halo" orbit, in which a spacecraft performs long, lazy curved loops around the libration point, almost as if the libration point were a celestial body.

A halo orbit is an N-body problem, involving interactions between the probe, the Sun, and the Earth. Early calculations of halo orbits involved a "successive approximation" approach: making educated guesses, running them through a software simulation, and then using the result to get a better guess.

This method was workable but crude. In the 1990s, improved software was developed that essentially mapped out entire ranges of three-body trajectories, known as "manifolds". Paths known as "dynamical channels" can be identified on the manifolds that chart out the course a spacecraft would follow on its own after given an initial push, analogous to the way a ball bearing would meander about on an uneven surface after being given a nudge one way or another.

The Genesis spacecraft, for example, was initially launched into a parking orbit around Earth. A short engine burn then sent it drifting towards the Sun, and three months later another brief burn put into a halo orbit around the L1 position. It then performed a total of four halo orbits, each lasting six months. After the end of the sampling mission, another burn sent the probe on a orbit that brought it all the way back to a halo orbit around L2, where MAP was making its measurements, and then to a close Earth flyby, where Genesis dropped its sample package.

The reason for this roundabout maneuver was to allow the package to be ejected during daylight hours. Since it was to be recovered by helicopter to prevent contamination, it could not be sent back at night. Unfortunately, although the elaborate flight plan went as intended, the capsule did not deploy its parachute after reentry in September 2004 and crashed into the desert. Some of the payload proved recoverable. Martin Lo, one of the leaders of the Genesis programme, called the elaborate network of manifolds snaking throughout the Solar System the "Interplanetary Superhighway". Jupiter and Saturn also have extensive sets of large moons that set up networks of manifolds on a smaller scale. By exploiting dynamical channels along these manifolds, the probe could maneuver from moon to moon, remaining at each for as long as desired, and then changing position with a slight nudge from its rocket engine.

UNCERTAINTY IN ORBITAL MECHANICS / THE BUTTERFLY EFFECT

- It might be thought from the examples of orbital mechanics shown so far that the motions of heavenly bodies follow a neat, predictable clockwork regularity. This was much the way things were seen in Lagrange's time, but in modern times this is a somewhat misleading impression. Tidy resonances and other orderly arrangements of celestial bodies will occur when relatively large forces can be focused on these bodies in a consistent fashion. For example, the Trojan asteroids are trapped in their orbits by the Sun and Jupiter, the biggest objects in the Solar System and with the most powerful gravity.

Such orderliness is not always the case. As the discussion of software for orbital mechanics shows, the gravitational interactions of all the bodies in the solar system are impossible to calculate with perfect precision, and approximate methods are required in practice. These methods are workable, of course, but they have limitations in their accuracy that can become important over the long term.

The best example is that of the orbits of most asteroids. A set of sightings can be used to obtain the orbit of an asteroid, but over time that orbit will shift in an unpredictable fashion, partly due to the unpredictable gravitational interactions of the asteroid with all the other bodies in the solar system, and partly due to other subtle effects, such as the pressure of sunlight. These little "nudges" on the orbit of an asteroid don't amount to much at any one time, but over a long period they can modify the asteroid's orbit to the point where it is "lost" and has to be "discovered" again.

- This uncertainty in orbital mechanics is a good example of uncertainties in physical systems in general. In an introductory document like this, the focus is on "linear" systems, in which the result of a set of influences on the system is directly proportional to the sum of the influences. Add masses to a box, the total mass of the system is just the sum of the individual masses.

However, physical systems may be "nonlinear", in that the result of the individual influences on the system don't add up in a nice uncomplicated way: a small shift in snowpack can trigger a huge avalanche in a "cascade" process. In the worst case, influences may add up in ways so that the results become impossible to predict over some period of time: influences too small to measure eventually grow and take the system off into some unpredictable direction. Such systems are known as "chaotic". The best-known example of a chaotic system is the weather. Everybody knows that unless the weather in a location is extremely predictable, the accuracy of official weather forecasts falls off drastically as the forecast is extended into the future. Forecasts are likely to be accurate for the next day, but they are very unlikely to be accurate for a week into the future.

Physicists like to describe the instability of the weather with the "butterfly effect", the idea that the movement of the wings of a butterfly in China one day might produce a tornado in Texas a week later. Forecasts can be improved with better computer simulations, throwing more computing horsepower at the problem, but that doesn't eliminate unpredictability, it just pushes it a bit farther out into the future.

Surprisingly, chaotic systems are not completely unpredictable. Once they are isolated from disruptions they may settle into patterns known as "attractors", in which the elements of the system move within certain limits, as if they were "attracted" to that range of paths. There are different specific classes of attractors, such as the "Lorenz attractor", a pattern of paths that looks like the eyes of an owl.

It was named after Edward Lorenz (born 1917), one of the pioneers of modern "chaos theory", who came up with the butterfly effect in the early 1960s. There is also the famous "strange attractor", which is a "fractal pattern". Fractals are geometric figures in which their overall appearance is duplicated at greater levels of "zoom" in on the figure. Natural examples of fractals are clouds, shorelines, and snowflakes.

Chaos theory was something of a scientific fad at one time, being inflated, often along with fractals, into a "new science", which was probably a bit of an exaggeration since the basic ideas have been around for a century or so though they have only been investigated in detail in recent decades.

Some particularly enthusiastic advocates even claimed that chaos theory reduced the traditional predictable linear mathematical models of physical systems to myths, but that was going much too far: the linear systems were always known to be approximations to reality, and they generally proved good enough approximations to get the job done as well as it needed to be.

This document uses the simplest, crudest possible mathematical models and gets a fair amount of mileage out of them.

On the other side of the coin, critics suggested that it was really the simple models that did all the useful work, and so chaos theory was mostly big talk that was of little practical use. However, if it wasn't the "new science" people claimed it was, it was still a valid element of systems theory and had its applications.

Once again, an understanding of nonlinear and chaotic systems is not in itself all that useful in a simple physics document. The issue is simply to warn against complacency: the real Universe is not always so simple. It was possible for Laplace to state that, given a knowledge of initial conditions and the rules of mechanics, the behaviour of the Universe could be predicted indefinitely into the future, but nobody believes that now.

Modern science has shown there is a fundamental uncertainty in physical measurements, below which it is literally impossible to obtain any more detail, and given nonlinear systems such undetectable variations may grow, in time, to a cosmic scale. Like the saying goes, science isn't necessarily an exact science.

FOOTNOTE: LARRY NIVEN'S INTEGRAL TREES / SPACE TETHERS

- Science-fiction writer Larry Niven wrote a pair of novels titled the integral trees (1984) and the smoke ring (1987) that have some interesting insights into orbital mechanics. These two novels postulated a double star system, consisting of a small white dwarf star named "Voy" in orbit around a bright central star named "T3".

Voy had a giant planet named "Gold" in orbit around it in turn. Voy had once been a bigger, brighter star itself, but the stellar explosion that turned it into a hot stellar cinder stripped Gold of its gaseous outer layers, resulting in a ring or "torus" of gas that stretched completely around Voy in the planet's orbit.

This "Smoke Ring" was long-lasting and contained enough debris to support the evolution and propagation of life. Plants evolved that produced oxygen, eventually making the Smoke Ring an environment where humans could breathe and survive. When a mutiny stranded an interstellar expedition from Earth there, successive generations of humans colonized the Smoke Ring and gradually spread all the way around it.

The major landmarks in the Smoke Ring were the "Integral Trees", which were huge free-floating trees about a hundred kilometers long that were curved into an elongated "S" shape, similar to the integral symbol used in calculus. Humans tribes colonized the tips of the Integral Trees, which were covered with jungle-like "tufts" of greenery, while the intermediate parts of the tree were generally barren.

The orbit of an object is defined at its centre of mass, which was at the centre of an Integral Tree. Since the Integral Tree was an elongated object, tidal forces from the white dwarf star forced the Integral Tree to point toward Voy at the centre of its orbit, crossways to the tree's motion.

Incidentally, this is known as "gravity gradient stabilization", and small satellites have been built with a mass on an extension to ensure that they always point straight down to the surface of the Earth as they orbit around it.

In any case, an independent object orbiting below or "in" inward to the centre of the orbit around Voy of the centre of mass of the Integral Tree would have a faster orbit than the tree, while an independent object orbiting above, or "out" outward relative to the orbit of the tree around Voy of the centre of mass of the Tree would have a slower orbit.

However, since the Integral Tree was a solid object, the entire Tree orbited at the same speed. This meant that the inner tuft of the Tree was moving slower than independent objects in the same orbital radius, and the outer tuft of the Tree was similarly moving faster than independent objects in the same orbital radius.

At the inner tuft, since the gases in the Smoke Ring were moving at their natural orbital speed but the tuft of the tree was moving more slowly, the tuft was in a perpetual howling wind blowing from behind the Tree's direction of motion. The inner tuft of the Tree curved in the direction of its motion to offer less resistance to this wind, while the tuft filtered debris from the wind to support the growth of the Tree. Gravity was no longer completely balanced by centripetal acceleration, and the result was that the humans felt a net gravitational force that grew stronger the farther "in" they went. Of course, they had to stay on the inside of the curve of the tuft, since if they went below they could fall down into the Smoke Ring.

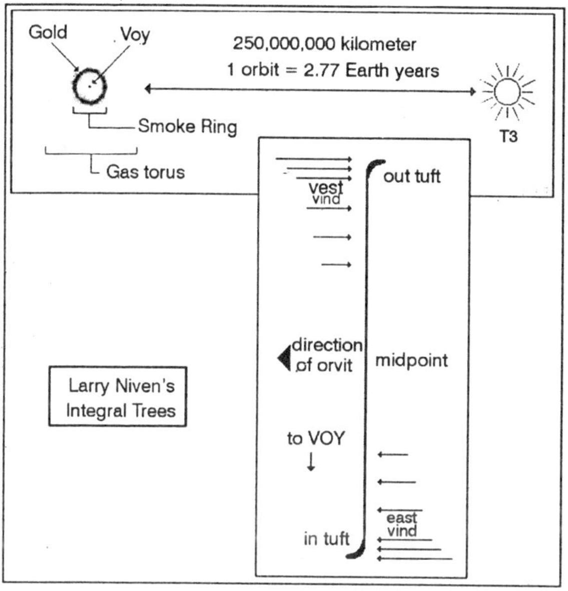

In a complementary fashion, the outer tuft was moving faster than the gases at its orbital radius, and so the outer tuft faced a similar howling wind but in the opposite direction, coming from in front of the Tree.

As with the inner tuft, it curved away from the wind and the tuft collected debris blown into it by the wind. Centripetal acceleration was no longer balanced by gravity, and so the humans felt a net "artificial gravity" that grew stronger the farther "out" on the tuft they went. Once again, they had to stay on the inside of the curve, or they would be flung up into the Smoke Ring by their excess velocity.

The excess gravitational pull on the inner tuft of course had to be balanced by the excess centripetal acceleration on the outer tuft, or the Integral Tree would not have been in a stable orbit. This set up a force between the two ends of the Tree that increased as the Tree grew longer, and so limited its length.

In fact, old or diseased Trees could snap apart at the middle, with disastrous consequences for the human colonies in the tufts. By the way, Smoke Ring materials unsurprisingly tended to accumulate in the L4 and L5 libration points defined by Voy and Gold, and these locations amount to "metropolitan areas" in the Smoke Ring.

The idea of making movies out of the integral trees and the smoke ring is appealing, though the special effects would be so difficult that an animated film might be the best approach, particularly since the human colonists had evolved into very tall and gangly forms in the low gravity and could not be played by any normal human actor.

Unfortunately, the environment of the Smoke Ring is so alien and disorienting that most viewers would not be able to figure out what was going on, and it seems plausible that if shown on the wide screen it might cause vertigo and motion sickness.

- Related concepts of orbital mechanics are embodied in some schemes for space tether systems. Tethers have already been described in an earlier chapter for setting up artificial gravity, but they have other potential applications. Imagine a space station in orbit around the Earth that has a tether hanging downward with the end not far above the Earth's atmosphere, and a second tether hanging upward from the station, with the end in a higher orbit.

Suppose the station crew wants to send a payload back down to Earth. The current way of doing this is to use retrorockets to kill the payload's orbital velocity, causing its orbit to decay.

Using a space tether, all that has to be done is winch the payload down the "down" tether. If the payload were free to orbit on its own it would speed up as it descended, but since it's linked to the station with the tether, it can't. Once it reaches the bottom of the tether it is released and, lacking the velocity to stay in orbit, reenters the Earth's atmosphere.

Of course, since the station slows down the payload as it is winched down the tether, the payload similarly makes the station speed up the tether connection is two-way, after all. That means that not only is the payload returned to Earth more cheaply than if retrorockets were used, but the station gets a "free" orbital reboost out of the transaction. Similarly, to send a payload to a higher orbit, the station crew can winch it out along the "up" tether.

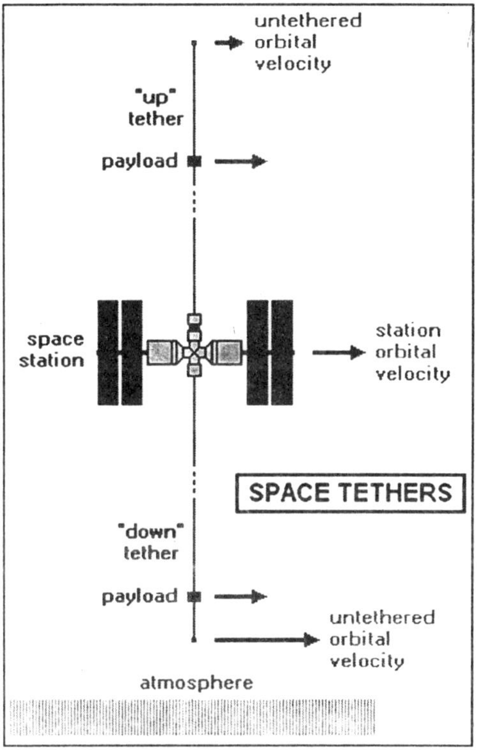

If the payload were free to orbit on its own, it would slow down, but it is dragged along with the station in its orbit. When the payload reaches the end of the tether, it is released and sails off outward into space. Once again, the tether works both ways: the payload slows down the station a bit, and when the payload is released the station falls to a lower orbit.

Of course, the "down" tether can be used to winch up payloads from low orbit, at the expense of lowering the station's orbit, and the "up" tether can be used to winch down payloads from high orbit, raising the station's orbit. This leads naturally to a more elaborate concept, the spinning tether or "rotavator".

Much like before, tethers are hung from the two sides of a central "hub", but in this case the hub is spinning and the ends of the tethers are going in great circles, falling towards Earth and then arcing upward into space again. The idea is that a rocket-boosted aircraft carrying a payload on its back might skip momentarily above the atmosphere, with the maneuver precisely timed so the rotavator sweeps down and snags the payload off the aircraft's back. The rotavator then curves around 180 degrees to release the payload into a high orbit. Another rotavator orbiting above the first might then snag the payload again from its new orbit and release it into a still higher orbit and so on. Rotavators could also be used to snag payloads from higher orbits and pull them down to lower orbits.

Since a rotavator loses a bit of orbital altitude with each payload sent upward and gains a bit with each payload sent downward, the rotavator scheme implies that the mass of traffic going up must be more or less matched by the mass going down.

There are other interesting things that can be done with space tethers, but they are beyond the scope of this document. So far, only small-scale space tether experiments have been performed, and the technology has yet to be put into practical application.

Chapter 6

Physics of Solids

- We are all familiar with the matter that makes up the world and ourselves, as well as the fact that it comes in three different forms: as solid, liquid, or gas. This chapter discusses the properties of solid matter, with a focus on the applications of those principles to the design of structures. Liquids and gases are discussed in the next chapter.

MATTER AND ATOMS

- The idea that matter could be broken down into submicroscopic individual units, known as "atoms", goes back to antiquity, though the concept didn't start to acquire scientific momentum until 1808, when the English scientist John Dalton proposed it to account for the results of his studies of chemical reactions.

Today, it is known that there are about a hundred different types of atoms with different weights and other properties. These "elements" include hydrogen, the lightest and most common element; carbon, which makes up soot; silicon, one of the constituents of sand; metals like gold, iron, and lead; the gaseous elements oxygen and nitrogen; and many others.

Atoms could also be combined to form "molecules". For example, if the element oxygen is burned with the element hydrogen, each oxygen atom forms "chemical bonds" with two hydrogen atoms to form water molecules. It was the distinct and predictable patterns of combinations of atoms into molecules that led Dalton to propose the atomic theory. Large numbers of atoms can combine into a single molecule, particularly in the case of the complex carbon-based "biomolecules" that make up living organisms, and the materials made from atoms and molecules can be arranged into large-scale systems, from stones to machines to flowers to human beings.

Detailed discussion of chemical reactions, in which atoms are converted into molecules or molecules are broken down into atoms, needs to be reserved for a document on chemistry. However, it is useful to comment here that such chemical reactions observe the fundamental law of energy conservation as obeyed in physics in general. Reactions may be "exothermic", releasing energy stored in chemical bonds, or "endothermic", absorbing energy stored into chemical bonds; but in any case the total energy before and after the reaction will remain the same. Only the forms are changed.

For example, water can be broken down into hydrogen and oxygen by running an electric current through it; this is an endothermic process. A spark will then cause the hydrogen and oxygen to explosively recombine; this is an exothermic process. On a broader canvas, a tree grows by using sunlight and raw materials such as water and carbon dioxide to drive its complicated biological processes, which in sum amounts to an endothermic process. The wood can be burned to release its energy, which is an exothermic process.

- Oddly, although atomic theory now seems to be an essential element of classical physics, it wasn't completely accepted by physicists until the early 1900s. Ironically, this was at the very time that the doctrines of classical physics would begin to be seriously challenged.

Atoms had been originally thought of as indivisible, but even before the final triumph of atomic theory physicists were beginning to realise that atoms themselves were not indivisible and had an internal structure. They gradually realized that an atom had a heavy central "nucleus" that had positive electric charge, surrounded by light orbiting "electrons" that had a negative electric charge. Electrons could be stripped from atoms to be sent through wires as electrical currents.

If an atom lost its normal complement of electrons, it became a positively charged "ion". Atoms could also acquire an excess of electrons, becoming negatively charged ions. Such charge imbalances were one mechanism that contributed to the bonding of atoms into molecules, though subtler processes were involved as well.

Physicists soon found that the interior of the atom was a complicated place where things happened that were impossible at a larger scale, and this simple model of a "classical atom" had to be revised. The revisions lead into the realm of modern physics and so are beyond the scope of this document. The classical atom still provides a useful basis for the discussion of the fundamental properties of matter.

GROSS PROPERTIES OF SOLIDS

- Solid matter covers a wide range of forms, everything from a compacted block of trash containing every form of household rubbish to a glittering diamond, consisting of carbon atoms arranged in an orderly, uniform "crystalline" arrangement.

At a high level of abstraction, chunks of matter have certain properties that can be characterized and measured, for example colour; density, or ratio of mass to volume; and various measures of strength, or the ability of that chunk of matter to retain its structure or "integrity" when subjected to various forces. Obviously these things are much easier to characterize for, say, a diamond than for a block of compacted trash.

Solid matter also has certain properties of "scale" that are independent of what the matter is made up of. For example, let's take a set of cubes of solid matter without regard to what the matter is. Suppose one cube is a centimeter on a side. Now let's suppose Dexter stacks up the cubes to make a bigger cube that is two centimeters on a side.

This doubles the linear dimensions of the cube. The surface area of the cube has expanded by a factor of four, while the volume and mass of the cube has expanded by a factor of eight. In more formal terms, doubling the linear dimensions of an object squares its surface area and cubes its volume and mass. This means that as an object grows bigger, its volume and mass increase much more rapidly than its surface area.

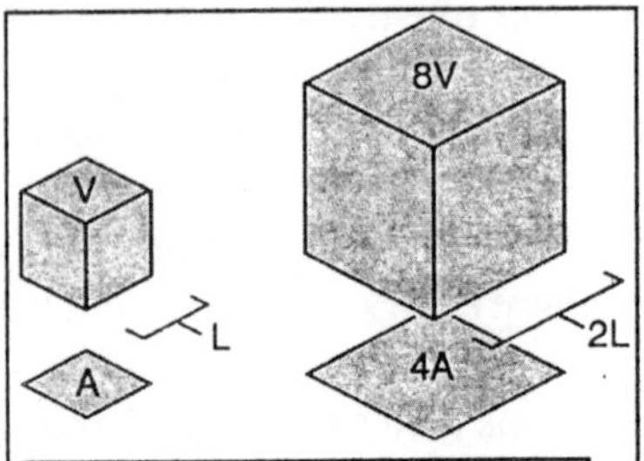

Scaling Properties of Bulk Matter

1950s horror movies such as Them envisioned monster ants, but in fact such scaling effects prevent insects from becoming very large. For one thing, they don't have lungs, and have to acquire oxygen through pores in their chitinous skins. Doubling their size would increase their volume and mass twice as fast as it would increase their surface area, halving their ability to obtain oxygen for that mass, and at a certain size a giant insect would simply suffocate.

A more important scaling issue that rules out giant insects is the issue of "compressive strength", or the ability of a structural support to bear weight placed on top of it. The compressive strength of a structural support, such as a column that holds up a building, is proportional to the cross-sectional area of a horizontal slice through the column.

A cross section is just one face of the surface of an object, and so like surface area increases with the square of a linear increase in size. However, the mass of the building increases with the cube of a linear increase in size.

Double the linear dimensions of a building and the ability of a column to support the building's weight is increased by four, while the mass and the load on the column is increased by eight. The column now has to proportionately bear twice as much load as it did at the smaller scale. This means that more, or disproportionately large, columns must be used to support a larger building. Insects generally have spindly bodies and legs, while an elephant has great stumpy massive legs. If an insect was scaled up to large size, it would simply collapse of its own weight. However, horror movie fans can still take comfort that deadly venomous or parasitic insects are not ruled out by the laws of physics.

STRENGTH OF MATERIALS

- These considerations of scale lead back to the notion of strength of materials, which in turn leads to consideration of how materials are used for building structures.

Although builders always had useful concepts of what could or could not be done when building structures, the first person to try to perform scientific experiments to determine the strength of materials was the Italian polymath Leonardo da Vinci. The first formal text on the subject of materials was actually published late in life by Galileo in 1638, titled dialogues concerning two new sciences. He was under house arrest at the time, having antagonized the Vatican by promoting the notion that the Earth went around the Sun and not the reverse, and materials seemed like a noncontroversial subject not likely to get him into further trouble.

Galileo's dialogues was not particularly correct in all details the basic laws of physics needed hadn't been invented just yet but the book did get people thinking. Isaac Newton had little interest in the subject of materials, but his rival Robert Hooke was fascinated by the subject, examining the ways in which materials responded to forces, in particular coming up with Hooke's law. This was mentioned in a previous chapter as:

$$\text{Force} = \text{stiffness} \times \text{compression}$$

It was stated then as applying to a spring, but it also applies to structural element made of hard metals, stone, and glass. It does not in general apply to flexible materials like rubber or tendon, which stretch easily up to a limit but then yield diminishing returns. Hooke's work was the first to at least imply the notions of "stress" and "strain". Stress is a force applied to a structural element. Such forces include:

- Compression forces, which squeeze the element in on itself.
- Tension forces, which try to pull the element apart.
- Shear forces, like we exert on a piece of cardboard to rip it in half.
- Torsion or twisting forces.

• Bending forces.

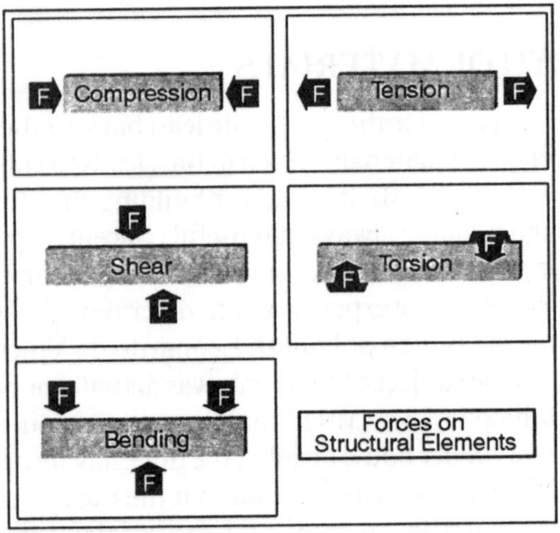

Stress is measured in terms of force per unit of area, or what is simply known as "pressure". The formal metric unit of pressure is the "pascal (pa)", which is a newton per square meter, named after the great French mathematician and physicist Blaise Pascal. Since a pascal is a somewhat small unit, it is often expressed in terms of "kilopascals (kpa)" thousands of pascals or "megapascals (Mpa)" megapascals. In compression and tension forces, the force is applied directly through a cross-section of a structural element, and so the stress is measured in terms of Mpa.

In shear force, the force is applied along a plane through a structural element, and so stress is measured in terms of Mpa over the area of the interface between the two sections of the structural element. Torsion is similar to shear but involves a torque over the area of the interface.

Bending is complicated, since it involves a shear force down through the structural element, a tension force along its bottom, and a compression force along its top. In practice, structural elements may be subjected to a number of these stresses at the same time. A structural element under stress will exhibit a "strain", that is, a lengthening, given in a simple proportion. Up to a certain limit, once the stress is removed, the structural member will return to its original length. This is called "elastic" behaviour". It was certainly implied by Hooke, but the idea was developed further by Thomas Young, who published a paper in 1807 in which he defined the concept of "elastic modulus" or "Young's modulus". In modern terms, this is the ratio of stress of a particular material to its strain:

elastic modulus = stress / strain

It is somewhat more intuitive to express this as:

strain = stress / elastic modulus

Since strain is a simple proportion, the units of elastic modulus are still Mpa. For materials that stretch easily. the elastic modulus has a low value; for example, rubber has an elastic modulus of 7. The stiffer the material, the higher the elastic modulus:

Wood (spruce)	11,000 Mpa
Ordinary glass	70,000 Mpa
Steel	201,000 Mpa
Diamond	1,200,000 Mpa

Given these definitions of stress and strain, then Hooke's law can be rephrased as follows:

force = stress × strain

Hooke's law also leads to the concept of "strain energy", or the potential energy contained in a strained structural element, which is similar to the eq4uation for energy in a stretched spring:

strain energy = volume ×(1/2) × stress × strain^2

Once the stress on a structural member exceeds a certain limit, the member will be permanently strained out of shape, and will break if the stress is increased. Not all materials respond to forces in such a nice graded way. They may not deform at all up to the level of force where they snap. Such materials are said to be "brittle". Glass is a good example, since any practical experience shows how difficult it is to bend a glass rod without breaking it.

Materials are said to have a "breaking strength" or "tensile strength", rated in the tension stress that causes them to snap. The concept goes back at least to Galileo, who performed tests in breaking strengths, though he didn't have the theoretical tools needed to describe it well. Typical tensile strengths include:

High strength steel	1,500 Mpa
Ordinary commercial steel	400 Mpa
Cast iron	~200 Mpa

Aluminum alloys	~300 Mpa
Hemp rope	80 Mpa
Bone	40 Mpa
Nylon fibre	1,000 Mpa
Kevlar fibre	2,700 Mpa

Not all materials respond to different forces in the same way. Although metals are strong in both compression and tension, stone is very strong in compression but is weak in tension. Wood can be split easily by shear forces applied along the line of its grain, but will resist larger shear forces applied at a right angle to its grain.

Forces acting consistently on a structural element over a long period of time may cause a slow deformation, or "creep", that can lead to failure even if the elastic limit is never exceeded. Similarly, a cycle of changing stresses repeated over and over may cause "fatigue" that will eventually lead to the failure of a structural element.

- Although Hooke got the ball rolling, the notions of stress and strain were not articulated in their modern form until 1822, in a document published by a by a French mathematician named Augustin Cauchy. It would take time for Cauchy's work to sink in, since he wrote in a theoretical style that was more or less foreign to engineers, but now his concepts are regarded as one of the major foundations of structural design.

The underlying idea of all structural design is balance of stresses. If the stresses imposed on a structural element exceed the strength of the element, it will collapse. The stresses are set up by what engineers refer to as "loads". Some of the loads on a structure are obvious: the "dead load" set up by the weight of the structure, and the "live load" set up by the people, furnishings, equipment, and materials contained in the structure. These are known as "static loads" because they don't change, or don't change much, over time.

Others loads are not quite so obvious. There are the "hidden" loads, with stresses set up by the expansion or contraction of structural elements due to changes in temperature, or by unequal settling of the building's foundations. Structures may also be subjected to rapidly changing "dynamic" loads, such as those caused by winds and earthquakes.

Dynamic loads can be particularly treacherous since they are somewhat unpredictable. In one famous example, the designer of the Citicorp skyscraper in New York City received a letter from a student who was curious about the building's stability in winds. The architect replied that the building's design had been analyzed to ensure that it would stand up to storm winds and that it was safe.

However, after sending his reply, the architect realized that the analysis had been performed with the assumption that the winds would be perpendicular to the faces of the building. He re-did his analysis to see what would happen if the winds were at an angle, and to his shock found out that a severe windstorm would generate torsion forces that would bring the skyscraper down in the middle of Manhattan. The Citicorp building's steel framework was hastily reinforced.

DESIGN OF STRUCTURES

- Structures can be divided into two classes: compressive structures, in which the structural elements are stacked on top of each other; and tensile or "suspension" structures, in which the structural elements are suspended from supports.

One of the simplest examples of a compression structure is a brick wall, with the bricks piled on top of each other. One of the simplest examples of a tensile structure is a circus tent, with the canvas "big top" held up by ropes hung from a central pole. Of course, in practice most structures feature both compressive and tensile elements.

The first really big structures were compression structures, since it was difficult to build durable large structures of wood, and stone doesn't lend itself to tensile structures. The major problem with stone compressive structures is that stone is very heavy, or more specifically has a low "strength to weight" ratio. This means that the weight of such a structure grows very quickly as it the structure grows bigger, and walls have to be made thicker and thicker at the base so the structure doesn't collapse. This approach ultimately leads to the simplest of all large stone structures, the pyramid, built not only in Egypt but in China and in the Pre-Columbian New World as well.

- Pyramids are an impressive but purely brute-force design. Builders were also learning more sophisticated tricks working with wood, the other classic building material. Wood was used to construct ordinary buildings, such as homes, and structures such as bridges.

The earliest wooden bridges no doubt consisted just of two beams set up across a stream and covered with planks. The major problem with this is that such a bridge is limited in length. It is supported at the ends but not in the middle, and as it gets longer so does the "lever arm" of the half-spans of the bridge, meaning that a person stepping on the middle of the bridge places a proportionately greater stress on it.

If the bridge was over the top of a shallow stream, this problem could be addressed by driving pilings in the middle of the stream and fastening the middle of the span to them, but if the stream wasn't shallow or if the bridge was over the top of deep canyon, such a fix wasn't possible.

There were two possible solutions. The first was the rope "suspension bridge", a tension-based design. Wooden or stone towers could be constructed on both sides of the canyon, with heavy ropes slung across their tops and anchored to the ground. A wooden walkway could then be suspended from the ropes. This type of bridge has been around for millennia, and is thoroughly familiar from B-fiction adventure movies since they are excellent sites for dramatic action scenes. They always seem to be in a frighteningly bad state of repair, and the waters beneath are often full of crocodiles, piranhas, or other predatory creatures.

For present purposes, a second option is more interesting. Suppose Dexter has a wooden bridge and wants to brace up the middle of the span from each end. He can do this by fitting a beam from the bank to the midspan at an angle. This allows the forces directed downward at midspan to be shunted towards the banks. Since this sets up a sideways force on the two support beams that tends to splay them apart from the middle, it is also useful to link them together at the bottom with a third beam, forming a triangular support structure known as a "truss". The truss is a basic structural form, and is composed of various triangular structural subunits linked together. The triangular shape is very useful as it will stand up to stresses imposed in different directions, and it is also employed in the "cross-bracing" of structures used to keep them from falling over in heavy winds and the like. The truss is of course in as wide use as it ever was; one of the tidiest examples is the metal high-voltage transmission line tower, which is a lightweight structure made up of little triangular truss elements.

- The problem with wooden bridges is that they didn't tend to last very long. Stone is far more durable, but it does not lend itself to construction of trusses. However, the same underlying principle of diverting stresses to the sides could be used on stone bridges as well, by building them with curved structures called "arches". The arch did imply the provision of supports to deal with the sideways loads it set up. This was generally simple with bridges, as they might be built between stony riverbanks or, if such weren't available, heavy footings could be constructed on each bank.

This could be a tricky matter for other structures. The Gothic cathedrals of Europe, with their high arched or "vaulted" ceilings, were built with "flying buttresses", structures attached to the outside of the building to provide the necessary lateral support. The "dome" was an extension of the arch, rotated around in a full circle. Like the arch, the dome required lateral support to handle the horizontal stresses. Islamic architecture, which demonstrates a certain fondness for domes, often involves half-domes or "semi-domes" ringing the main dome to distribute sideways forces down to the heavy walls of the buildings.

Although complete spheres are not ordinarily used in buildings, they are an extension of the notion of an arch or dome, in which compression is distributed around the sphere. An egg can be surprisingly hard to crack if subjected to balanced compression forces try carefully squeezing a hard-boiled egg and spheres are often used in the crew-carrying elements of deep-ocean submersibles, since they can withstand extreme pressures.

Most basic elements of classical building design, like the arch and dome, were in place by Roman times. Major advances in structural design had to wait until the development of large-scale metalworking in the 19th century. Iron and then steel structural elements, with their strength in both compression and tension and their high strength-to-weight ratio, made possible great steps forward in building design.

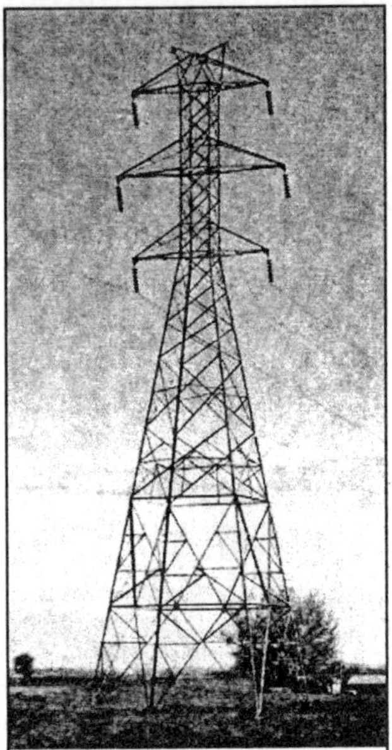

Powr Tower (GVG/PD)

Classical construction forms, such as trusses and arches, could be built with such metal beams. The beams were generally in the form of "I-beams", which consist of a length of metal with a long flat horizontal plate on the bottom, a similar long flat horizontal plate on top, and a vertical plate connecting the two horizontal plates. A beam is subjected to compression on top and tension on the bottom, and the I-beam is designed to provide the maximum cross-sectional area on bottom and top to resist these stresses.

Such techniques could be used to build new sorts of structures, such as the modern skyscraper. The poor strength-to-weight ratio of stone made very tall buildings impossible, but metal structural elements allowed buildings to be made much taller. A skyscraper consists of a "cage" of steel girders that supports the floors and the walls.

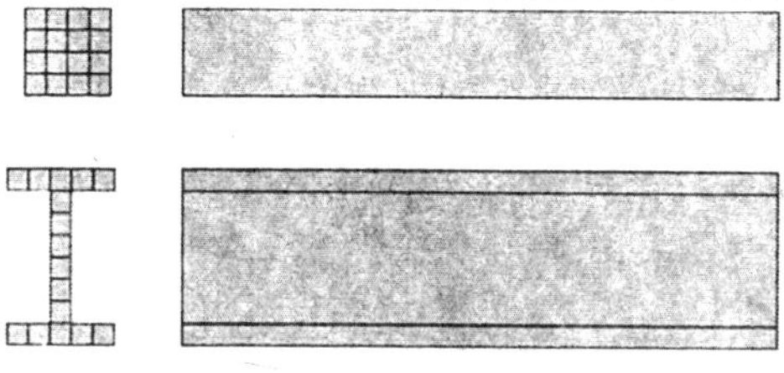

SIMPLE BEAM VERSUS I-BEAM

The spectacular explosive demolition of such structures, which usually makes local or even national TV news, is based on the sequential detonation of charges that cut critical structural supports of the cage in a particular order, causing the building to cave in on itself in a heap if the event has been properly planned.

Metal building elements also allowed the development of large truss bridges, with a latticework of metal beams to allow distribution of stresses. It also led to the development of modern suspension bridges in which the roadbed is hung from huge cables slung over high towers and anchored to heavy concrete "anchors" at each end and "cable-stay" bridges in which the roadbed is held up by cables connected directly to the towers.

Concrete was around in Roman times, though the recipe was lost in the Middle Ages and not rediscovered until the 19th century. Modern building techniques make it far more useful than it was to the Romans. Using steel reinforcing rods, it is commonly used for foundations, walls, and other structural elements under compressive loads.

However, "prestressed" concrete elements can also be used to handle tension loads that would cause ordinary reinforced concrete to crack. A prestressed concrete structural element is cast in a frame that supports steel cables under tension and laced through the concrete. When the concrete hardens or "cures", the frame is removed, and the tension of the cables within the concrete matrix prevents it from cracking under tension loads.

- There is a balance in the design of structures in the

number of structural elements required. Reducing the number of elements reduces complexity, generally reducing the labour of assembly and cost. However, reducing the number of elements also usually means larger elements, which may be difficult to handle, and also more prone to damage.

Suppose a dome is made up of curved panels; the fewer the number of panels, the larger the panels, the greater the load each panel has to carry, making each panel more prone to cracking and also making replacement of the panels more troublesome. Of course, if there are a very large number of small panels, installing them will be laborious, and trying to make sure the dome doesn't leak in the rain through the seams between the panels may be difficult. There are cases where the benefits of "subdivision" are obvious. The cables used to hold up a suspension bridge are each made of large numbers of individual strands; the breakage of a few strands has little effect on the strength of the cable as a whole. It is also possible to construct structural elements as "laminates", using layers of materials that are bonded together with adhesives.

Plywood is a common example of a laminate, and panel structures may be made as "sandwiches", with a lightweight core and stiff surface paneling, providing an improved strength to weight ratio. Simple corrugated cardboard is a common example of such a sandwich material; much the same approach can be used with metal panels as well, with a corrugated or honeycomb fill. However, laminates do present the danger of becoming separated or "delaminated" under stress.

Similar principles applies to the types of "composite" materials that consist of threads of one material embedded in a matrix of another material. Examples include glass or graphite fibers embedded in plastic, or "whiskers" of silicon carbide embedded in metal. Any failure of a few threads does not do much to undermine the integrity of the structural element made of the composite, and the threads prevent the propagation of cracks. Research is now being conducted on also incorporating sensors in composites in the form of threads to give indication of when a structural element is about to fail, but such "smart" materials are not in common use yet.

Chapter 7

Physics of Liquids and Gases

- This chapter follows up the discussion of the physics of solids by providing a survey of the physics of liquids and gases.

LIQUIDS: BUOYANCY

- Liquids have some properties similar to those of solids, such as colour and density, but of course they differ in their ability to flow and to take the shape of any container they are poured into. They possess "viscosity", which is a measure of their ability to flow. A "thick" liquid that is hard to shove through a pipe has a high viscosity, while a "runny" liquid that flows quickly through a pipe has a low viscosity.
- It is obvious that it is much easier to get vinegar out of a squeeze bottle than honey. Viscosity is related to the concept of "fluid friction": concepts of friction and drag apply to liquids, and a liquid with a high viscosity generates a high level of fluid friction.

Another obvious characteristic of liquids is that a body of liquid will seek a common level. A pond undisturbed by wind may be "smooth as glass", as the saying goes, and it will be perfectly level, at least if it's small relative to the curvature of the Earth. It has to be noted here that in more general terms, the angle of the surface of a liquid will be at a right angle to the direction of the force applied to a liquid, so water in a spinning bucket will tend to "climb up the walls" of the bucket. An undisturbed pond will be flat because it is only subject to the pull of gravity, which is straight down.

If pipes of various diameters are stuck through the surface of this flat pond, it only takes a moment's thought to realise that the water level is the same no matter the diameter of a pipe.

Yet another obvious property of liquids is that not only can they flow through pipes, objects can flow through or over them. This leads to the concept of "buoyancy", the tendency of objects to float or sink in a liquid. This is unsurprisingly due to the density of the object relative to the liquid. A block of wood is less dense than water and floats on water, while a rock is more dense than water and sinks.

The sunken rock will displace a volume of water equal to the volume of the rock. The floating block of wood has a slightly subtler behaviour. It will sink into the water under the influence of gravity until it displaces a volume of water equal to its own weight. At that point, it will reach an equilibrium of forces and sink no further.

This is known as "Archimedes' Principle", since it was devised by the Hellenic scientist Archimedes (287:212 BC). The forces supporting this floating block of wood will focus upward and inward on it, summing to a point known as the "centre of buoyancy". The centre of buoyancy is always above the centre of mass of a floating object; the farther the distance between the centre of buoyancy and the centre of mass, the more stable the object that is, the less its tendency to capsize.

According to tradition, Archimedes discovered this fact while getting into a bathtub, and used it to determine for the authorities if a crown being sold by a metalsmith was actually made of gold or a debased metal. Archimedes weighed the crown, then determined its volume from the amount of water displaced, and from that determined its density, which could be compared with that of pure gold.

The story says the metalsmith was found guilty and executed, but the tale is probably a myth: it is a bit unlikely that Archimedes, using the instruments available to him, could have performed the required measurements with enough accuracy to have detected a fraud unless the debasement was so gross that it would have been obvious by simply inspecting the crown. One simple example of the practical application of Archimedes' principle is the "Plimsoll line", which is a marking painted on the hull of a vessel in the form of a circle or diamond cut by a horizontal line. If the ship is sitting so low that the Plimsoll line is underwater, the ship is overloaded.

Another interesting example is provided by the question: if all the ice at the North Pole melted, how much would the sea level rise? From a simple consideration of buoyancy, the answer is: not at all. The polar ice displaces a mass of water equivalent to itself, and so when it melts, there is no overall change in the water level.

Of course, the ice melting off of Greenland and Antarctica would raise the sea level, and there's also the consideration of the extremely unpredictable ways that an ice-free Arctic Sea might affect global climate.

Incidentally, if a more dense liquid is poured into a less dense liquid, as long as the two liquids don't have a tendency to mix, the two will separate, with the less dense liquid on top and the more dense liquid on bottom. The classic example of this is oil and water.

- Considerations of buoyancy govern the operation of a submarine. A submarine cruising on the surface of the ocean has an average density somewhat less than that of water. To submerge, it has to take in water in "ballast tanks", increasing its density to more than that of water and causing it to sink. To rise again to the surface, it has to purge the ballast tanks by blowing air into them, forcing the water out and lowering the submarine's density.

Once underwater, a submarine needs to maintain a density comparable to that of water so that it will not tend to either rise or sink. This issue is somewhat complicated by the fact that the submarine's hull is compressed as the vessel dives deeper, reducing the volume of the hull and so increasing the submarine's density. In any case, once the submarine reaches the desired cruise depth, it can then maneuver freely by essentially "flying" underwater, using the "diving planes" or "foreplanes" that are mounted on nose or the "sail" (what's sometimes called the "conning tower") of the submarine, along with the "tailplanes", which are generally arranged in a vertical cross pattern.

However, there is an added complication for a submarine

moving underwater. It is not firmly supported, and so if its centre of gravity shifts from the middle of the vessel, it will tend to tip down or up, making it difficult to control. This issue, known as "trim", was a particular problem for early submarines. It was solved by adding an additional set of "trim tanks", in addition to the ballast tanks.

There is a trim tank in the forward section of the submarine and a trim tank in the rear section of the submarine, with trim maintained by adjusting the relative amounts of water in these tanks.

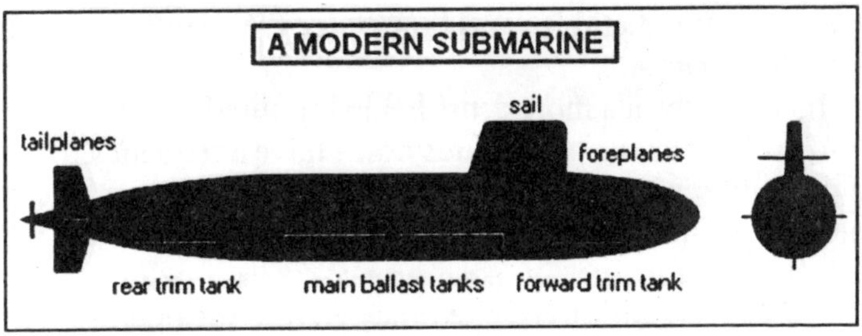

It takes a bit of time to pump water to shift trim, and so in World War II submarine warfare movies there is often a scene where the sub has to crash-dive, and all the crew that can be spared dash forward in a mad fire drill through the hatches to make the submarine nose-heavy.

LIQUIDS: FLUID PRESSURE / SURFACE TENSION

- Although it is not intuitively obvious, the buoyancy of the submarine does not change in any significant way, no matter how deep it goes. Water is essentially incompressible and its density is effectively the same at any depth. Drop an iron block into the ocean and it will continue to sink at its terminal velocity a concept that applies in liquids just as it does in gases all the way to the bottom.

However, there is a limit to how deep a submarine can go, which is imposed by "water pressure", or the weight of the water above the submarine, which increases linearly with depth.

Doubling the depth doubles the pressure. There's nothing particularly tricky about this idea. Suppose Dexter has a set of identical concrete blocks, and proceed to stack them up one by one on top of each other.

Of course, the force per unit area on the bottom of the stack grows linearly with each block that is added. There is no difference if he stacks up containers of water instead of concrete blocks, or for that matter adds uniform increments of water to a single tall container.

What is not quite so intuitive is that the water pressure is the same in all directions. Suppose Dexter applies pressure to the top of a tank of water. Then, ignoring the increment of pressure added by the water in the tank as it gets deeper, the pressure of the water against the sides and bottom of the tank is exactly the same as the pressure on the water on top of the tank. This is known as "Pascal's principle", Pascal having published the concept in 1647.

One way to understand Pascal's principle is to understand the operation of a hydraulic jack. Suppose Dexter has a tank filled with water, with two tubes containing pistons fitted into the top. One tube has a cross-sectional area of 10 square centimeters, while the other has a cross-sectional area of 100 square centimeters. If he exerts a downward force on the piston in the small tube, the transmission of water pressure on the piston in the big tube will result in an upward force ten times greater.

In other words, this arrangement gives Dexter a mechanical advantage of 10:1. As with levers and screw jacks and so on, of course he doesn't get something for nothing. A simple consideration of the relative volume of water moved from the small tube to the big tube in this example shows that the piston in the big tube only moves a tenth of the distance upward that Dexter shoves the piston in the small tube down.

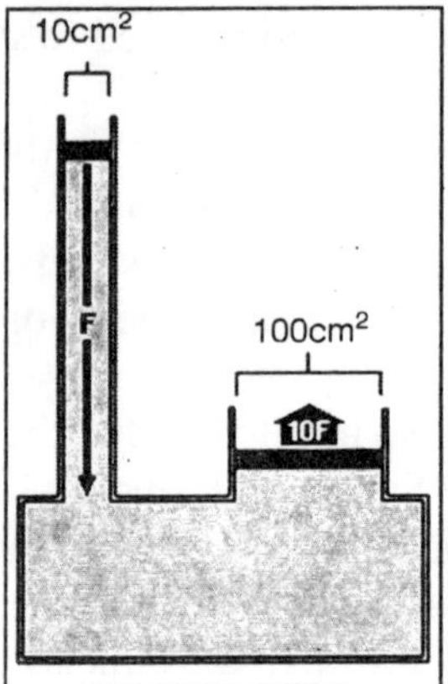

Hydraulic Jack

Another aspect of Pascal's principle is that it accounts for the buoyancy of a floating object, as discussed in the previous section. It is the force of pressure, exerted upward against the floating object, that prevents it from sinking further. Once the volume of water displaced equals the weight of the object, the force of water pressure will then support that object.

- In the definition of Pascal's principle above, the assumptions were that the surface of the tank was under heavy pressure and the tank was not very deep, meaning the pressure was the almost the same at the top and bottom of the tank and the difference could be ignored.

For water of any real depth, the increment of pressure caused by the accumulated weight of water as the depth increases is of course not negligible. Water has a mass of a tonne (1,000 kilograms) per cubic meter, and so on Earth, where the gravitational acceleration is 9.81 meters per second, a cubic tank of water a meter on a side exerts a force of 1,000

- 9.81 = 9,810 newtons on the square-meter base, or 9.81 kilopascals. If Dexter keeps the sides of the base the same size, but doubles the height to two meters, the pressure grows linearly to 19.6 kilopascals. If he goes up to three meters, the pressure goes up to 29.4 kilopascals. The water pressure on the sides of the tank increases linearly from zero at the top to the same as the bottom when it reaches the bottom.

However, let's suppose Dexter has a tank of water that's a meter high, a meter wide, and three meters long. This tank has a volume of three cubic meters. Is the bottom then under 29.4 kilopascals of pressure?

No, of course not. Pressure is force per unit area. He's increased the volume by a factor of three, but he's also increased the area of the base by a factor of three, and so the pressure remains the same as it was for the one-cubic-meter tank. The pressure is due to the weight of water pressing down only in the vertical direction. This means that it doesn't matter what size the base of the tank is. As long as the water in it is a meter deep, the pressure on the bottom in 9.8 kilopascals. This is more or less intuitive.

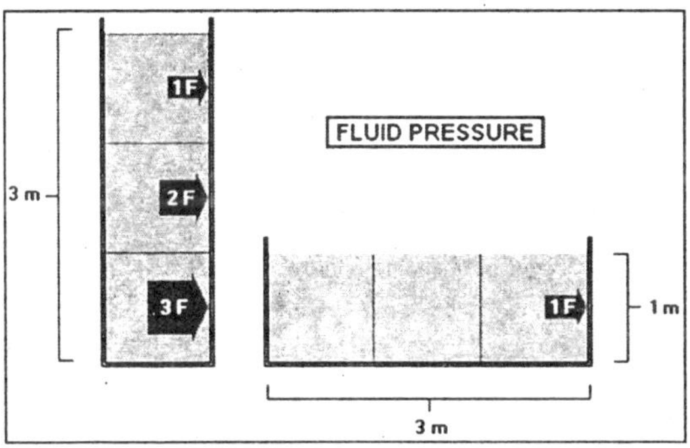

What is not so intuitive is that, as dictated by Pascal's principle, water pressure is exerted in *all* directions, and so a dam has to be just as strong to hold back a body of water that extends, say, only 10 meters behind the dam, as it must be to hold back a body of water of equal depth that extends 100 kilometers behind the dam. The depth is the only real determinant of fluid pressure.

There are two basic forms of dams: "gravity" dams, which simply use great mass to hold back the water, and "arch" dams, which are built in narrow rocky canyons in the form of a horizontal arch, which transfers the massive load of the water to the canyon walls. Some dams may use both mass and an arched configuration to ensure structural stability.

- Obviously, the pressure of a liquid is equivalent to potential energy. If Dexter installs a valve at the bottom of a tank and opens it, then the water will pour out in a stream and will be able to do work, for example by turning a water-wheel.

This should not be surprising. A solid mass held at a height has potential energy due to its elevated position in a gravitational field, and the only difference between that and a liquid is that the liquid is in effect a continuous series of masses held at a continuous range of heights.

This makes calculation of the precise value of potential energy of the liquid in a tank a bit tricky. We won't do that here, but the important thing is to realise that water pressure is equivalent to potential energy, in the same way that the tension of a compressed spring is equivalent to potential energy.

Now let's suppose Dexter decides to tap this potential energy by opening the valve on the bottom of the tank. Water floods out in a stream. Since the water is now moving, it has kinetic energy. Of course, the higher the pressure at the valve, or equivalent the greater the height or "head" of water above the valve, the greater the kinetic energy of the flow and the greater its velocity. This is known as "Torricelli's principle", after the Italian physicist Evangelista Torricelli.

Since conservation of energy says that Dexter can't get something for nothing, this increase in kinetic energy must be obtained by a decrease in potential energy. As water pressure is equivalent to the potential energy of a liquid, this means that the pressure must decrease as well.

In other words, the faster a liquid moves, somewhat counterintuitively the more its pressure decreases. However, it's not so different in concept from Dexter firing a ball out of a spring-loaded toy gun: the spring tension is released and converted into the kinetic energy of the ball. The same effect is observed when water or another liquid passes from a pipe with a wide cross-section into a pipe with a narrow cross section. For a given rate of liquid flow, measured in liters per second, the velocity of the flow, measured in meters per second, has to increase if the pipe diameter decreases to maintain the flow. In mathematical terms:

Liquid flow rate =
Cross section1 × velocity1 =
Cross section2 × velocity2

The increased velocity of the liquid flow in the narrow section of pipe will result in a decrease in liquid pressure. The following illustration demonstrates this principle, which is known as the "Bernoulli effect", after the Swiss physicist and mathematician Daniel Bernoulli.

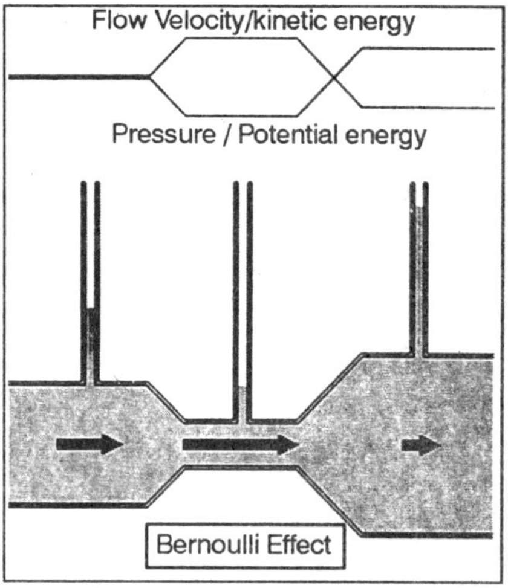

This illustration shows the pipe as horizontal. Of course, if it weren't, the liquid would flow "downhill" and pick up additional kinetic energy by the simple act of falling under gravity, complicating the analysis.

- On a small scale, liquids can demonstrate behaviour that would seem bizarre on a larger scale, due to attractions between the molecules of the liquid.
  For example, people can't walk on water, but insects like water-striders can. The intramolecular forces set up by water create "surface tension" that prevents a light object from simply falling right through the surface, even if it's denser than water.
  Similarly, the adhesion of water to, say, glass surfaces will cause it to crawl up into glass tubes stuck into the surface of the water.

This sort of "capillary action" is greater for tubes with very small diameters than it is for tubes with very large diameters. Again, this is due to a scaling effect, since doubling the diameter of a tube only doubles its circumference and the area of its glass walls, while the cross-sectional area and the amount of water that is to be lifted increases by four times.

One interesting application of these phenomena is the "restartable" rocket engine. A rocket engine that has to be restarted in space has to the deal with the problem that in free-fall the propellants won't flow down out of the tank to the engine, since there is no "down".

There are various ways to deal with this problem; one ingenious scheme relevant to the current discussion is to fit the fuel tank with a set of internal grids of finer and finer spacing towards the bottom of the tank. The grids set up a "wicking" action that pulls the propellant to the bottom of the tank.

GASES

- Gases have much in common with liquids. They can flow, for example, and less dense objects will float. Gases also are subject to considerations of pressure like to those of a liquid; the weight of the atmosphere presses down on those who live at the bottom of it, and this atmospheric pressure drops off as we go to higher ground or fly up into the sky.

The atmospheric pressure of the Earth at sea level makes a convenient reference point for discussions of pressure, and one older measure of pressure is the "atmosphere", or 101,325 pascals. This is a somewhat arbitrary measure of pressure, however, and not convenient to work with using metric units, but the value of 100,000 pascals, known as a "bar", can be used as a reasonable approximation, all the more so because the pressure of air at sea level actually varies with changes in humidity, temperature, and wind conditions.

A "barometer" is used to measure atmospheric pressure. A traditional scheme for building a barometer involves a U-shaped tube partially filled with mercury. One end of the tube is sealed and the other is open to the atmosphere. Mercury is squirted into the sealed end of the tube while it's upside down, and when the tube is turned over, the mercury falls down to the bend in the tube, leaving behind a vacuum in the sealed side.

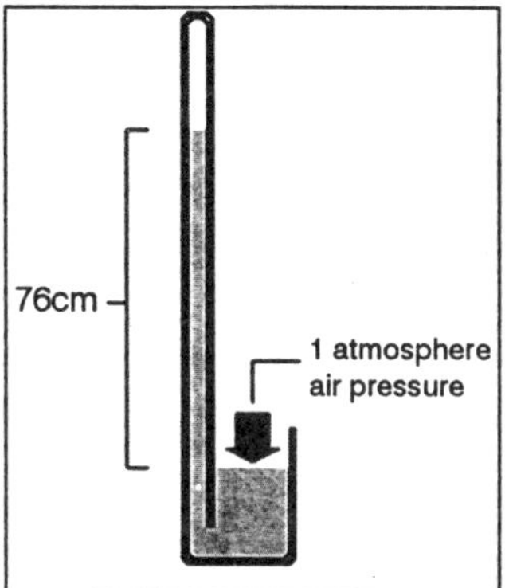

Mercury Barometer

Air pressure will push the mercury up into the vacuum, and the height of mercury on the sealed side will provide an indication of atmospheric pressure. The mercury will be 76 centimeters higher in the sealed side of the tube than the open side of the tube at a standard atmosphere of pressure. Of course, since pressure is force per unit area, it doesn't matter how wide the tube is, as long as its diameter is constant.

This scheme was discovered by Torricelli in 1643, and a now-disused measure of pressure named the "torr", equivalent to the displacement of a millimeter of mercury under standard conditions, was defined in his honour. Naturally, a standard atmosphere of pressure is equivalent to 760 torr, or 76 centimeters of mercury.

A more modern approach is to use a metal bellows that has been evacuated. The bellows is connected to an indicator needle through some mechanical arrangement that amplifies any minor motion in the bellows, and as the bellows flexes under changes in pressure, the needle moves to give the pressure value on the indicator scale.

This device is known as an "anaeroid barometer". If the indicator scale is calibrated to give altitude instead of air pressure, the device is called an "altimeter".

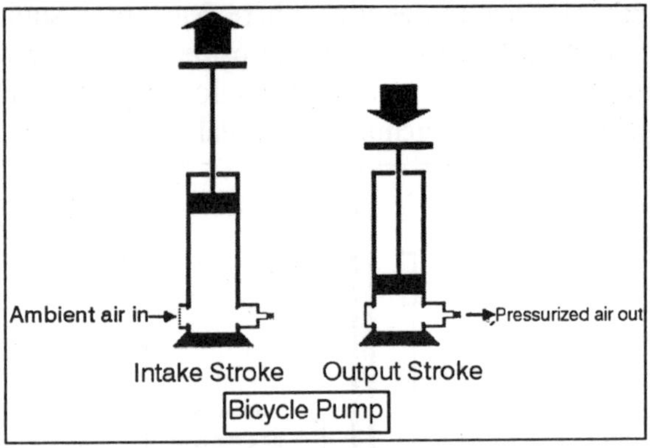

Manipulating the pressure of air or other gases is the basis of "pneumatic" technologies. The most basic pneumatic device is the hand-driven air pump or "bicycle pump" as it's sometimes called, which involves a cylinder with a piston that is moved up and down by muscle power.

There's two valves in the cylinder, an "input" valve that opens inward under a pressure difference and an "output" valve that opens outward under a pressure difference. Pulling up the handle on the pump creates lower pressure inside the cylinder, which causes the input valve to pop open, with ambient air entering to eliminate the pressure difference. Pushing down on the handle creates higher pressure inside the cylinder, forcing the input valve closed and popping the output valve open to push air into the tire or whatever.

- The notion of the mercury barometer leads to the concept of the "siphon", in which a liquid such as water can be driven from a reservoir to an outlet through a pipeline over a hill or other obstruction. This vertical diversion is called a "siphon", and it works as long as two conditions are obeyed: the outlet has to be lower than the surface of the reservoir, or in other words water will still only flow downhill overall, and the siphon has to be filled with water before the scheme will work.

This looks like a problem in hydraulics, and it is, but the principle behind it is the same as that behind the mercury thermometer: the column of water flowing up the siphon is supported by the air pressure pressing down on the surface of the reservoir. When water flows out the outlet side of the siphon, it creates lower pressure, and so water on the inlet side of the siphon is driven upward.

That's why the siphon has to be filled before the siphon effect will work: if the outlet side of the siphon was open to the air, the water level in the input side of the siphon would be exactly the same as it was in the reservoir.

There is a limit to the height or "crest" of the siphon above the reservoir level, depending on the density of the liquid and the air pressure above the reservoir. As mentioned, one atmosphere of pressure will lift mercury to a height of 76 centimeters, and since mercury is about 13.5 times more dense than water, one atmosphere of pressure on a reservoir should lift water about $13.5 \times 0.76 = 10.26$ meters.

This seems like a shallow crest for getting water over the top of a tall hill, but it must be remembered that the crest is relative to the water level in the reservoir, not relative to the bottom of the hill. Water reservoirs are often sited on high ground; the pipeline from the reservoir may drop down into a valley and then back up over a hill, but the crest of the hill will be about the same height or lower than the reservoir level.

- Buoyancy issues are as relevant in gases as they are in liquids. For example, if a balloon is filled with a light gas such as hydrogen or helium, it will have a lower density than the ambient atmosphere and tend to float up.

There is also the phenomenon of "lift" in heavier-than-air machines like aircraft and helicopters. There are misconceptions on how lift actually works, the biggest being that it is due to the Bernoulli effect, an error that actually pops up in elementary physics books.

As the story goes, the wings of an aircraft have a curved cross section, with a greater curvature on the top than on the bottom. Air flowing over the top of the wing has a longer path, and so its flow over the wing is more rapid. This means that the pressure on top of the wing is less than that on the bottom, and the effect is to provide a net upward force known as "lift" that keeps the machine flying. The problem with this scenario become immediately obvious when the question is asked: "So how can aircraft fly upside down?" They don't normally, since it's uncomfortable for the pilot, but any good flight demonstration team like the US Air Force Thunderbirds flies their jet fighters upside-down as a regular practice and the machines aren't immediately shoved into the ground.

The fallacy with the Bernoulli effect argument is that it assumes the airflow over the top of the wing has to maintain mass flow with the airflow over the bottom of the wing when it isn't constrained to do so. The Bernoulli effect may have some effect on lift, but it's not the primary mechanism for the phenomenon.

The actual reason for lift is Newton's third law of motion. In flight, a wing is held at an "angle of attack (AOA)" upward of the direction of airflow, with the airflow striking the bottom surface and being deflected downward as "downwash".

The downward velocity component of the downwash provides a reaction to force the wing (and the aircraft connected to it) upward. The need to maintain AOA relative to the airflow does not mean the aircraft has to fly nose-up all the time; the wing may be fixed to the fuselage at a predetermined AOA relative to the aircraft's flight axis.

As far as this description goes, a simple flat wing would work as well as any other, so why is a wing typically built with a curved upper surface? The trick is that such a wing would produce a good deal of turbulence over the top of the wing, causing "buffeting" as well as some "upwash" of the airflow, which would tend to force the aircraft back down.

Due to some subtle features of fluid flow discussed in a bit more detail below the airflow tends to hug the curve of the upper surface, which means that it leaves the trailing edge of the wing in a downward direction, also contributing to lift. The higher the AOA of the wing, incidentally, the greater the drag, and so there's a tradeoff between AOA and speed. It is not unusual for aircraft to have a landing gear configuration that leaves them nose-high, increasing the AOA until the aircraft gets off the ground and starts flying level. A high AOA is particularly important for carrier takeoffs and landings, since the aircraft has to get up or down in a short space.

A number of British carrier jets of the 1950s and 1960s were actually launched off carrier catapults strapped down so their nosewheel was in the air to ensure a high AOA. The US Navy Vought F-8 Crusader fighter had a wing that hinged upward for takeoffs and landings to permit a high effective AOA while giving the pilot a good view forward.

Such considerations of lift show that a conventional aircraft, even though it can in principle fly upside down, isn't as efficient in flight when doing so, having to maintain a nose-up attitude to ensure a usable AOA. Stunter aircraft may have no fixed AOA and wings that are curved on both sides.

- An aircraft has "flight control surfaces" that modify the airflow to control the machine's direction. There are moving panels at the trailing edge of each wing known as "ailerons" that are pivoted in an alternating fashion on one side up, on the other down to allow the aircraft to "roll".
- There are "elevators" at the trailing edge of the horizontal tailplane that are pivoted together to cause the aircraft to "pitch" up and down. Finally, there is a "rudder" at the trailing edge of the vertical tailplane that can be pivoted to cause the aircraft to "yaw" back and forth.

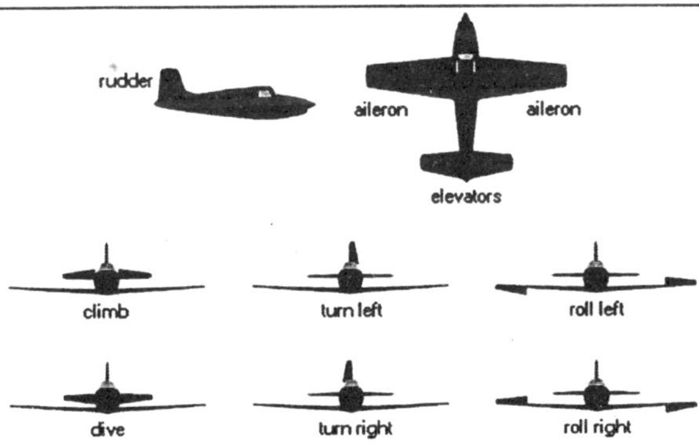

AIRCRAFT FLIGHT CONTROLS

These are the "basic" control surfaces, but there are many variations. For example, many tailless aircraft, usually delta-winged machines, have an "elevon" on each wing, with both moved up or down to provide elevator pitch control, or moved in opposite directions to provide roll control.

Tailless flying wing aircraft that don't have a tailfin, such as the B-2 Stealth bomber, have "rudderons" that are split top and bottom on the outer rear edge of the wing.

Both rudderons can be opened up top and bottom on one side to cause the aircraft to turn, or can be opened up top and bottom on both sides of the aircraft to act as airbrakes. An elevon is fitted on each half of the wing inboard of the rudderon to provide pitch and roll control.

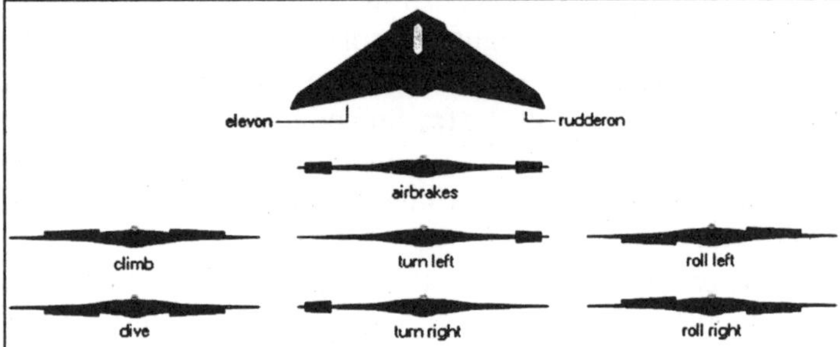

FLYING WING CONTROL SYSTEM

Incidentally, moving the rudderons away from their neutral position greatly increases the B-2's visibility to radar, so when it is flying into a combat area it steers by adjusting engine thrust, though this is more troublesome and less effective than using rudderons.

Aircraft may also have a number of auxiliary control surfaces. "Flaps" can be extended below the wings to increase the downwash and lift for take-off or low-speed flight, with flaps cascaded in some cases to increase their length. Control surfaces named "spoilers" or "lift dumpers" can be attached to the top of the wing to disrupt airflow and destroy lift, helping the aircraft to land. High-altitude aircraft have very wide wingspans and tend to be reluctant to come down on the runway, so their wings are usually fitted with lift dumpers.

Like a submarine, an aircraft has to maintain trim. This is obvious in the case of cargoes, which clearly can't be loaded onto a cargo plane in an unbalanced fashion, and the crew of a large cargo aircraft will generally include a specialist called a "loadmaster" who ensures that the cargoes are balanced properly incidentally, maritime cargo vessels also have loadmasters to reduce the hazard of capsizing due to an unbalanced load. Aircraft trim is less obvious in the important case of fuel management.

Usually aircraft designers will try to put fuel tanks at the centre of gravity of an aircraft, but an aircraft may have several fuel tanks, and care often must be taken to empty the tanks in a specific order, or the aircraft may become unbalanced. This was done manually by the pilot in the past, but now sophisticated aircraft have automatic systems to ensure the proper sequencing.

Since an aircraft will always have slight imbalances, instead of having the pilot try to compensate for trim faults by keeping the control surfaces slightly activated at all times, which would be very tiring, traditionally aircraft have featured small control surfaces called "trim tabs" that can be set to a fixed position using a dashboard knob to keep the machine flying level. Some aircraft may instead use knobs to introduce a slight bias in the aircraft's flight control surfaces.

COMPLICATIONS: VISCOUS FLOWS AND COMPRESSIBLE FLOWS

- The discussion to this point on the flows of liquids and gases which can be lumped together as "fluids", meaning both can flow has been simplified by assuming the flows aren't viscous that is, subject to friction forces and are incompressible. These are reasonable assumptions for a simple analysis, but the real world can be substantially more complicated.

Formal studies of the effects of viscosity on fluid flow began in the middle of the 19th century, leading to the invention of equations for describing the viscous flow of an incompressible liquid; the equations were developed independently by Claude Louis Marie Navier, a French engineer, and Sir George Gabriel Stokes, a British mathematician, and so these rules are known as the "Navier-Stokes equations".

They're not simple equations, and in fact they are only fully solvable for simple scenarios. Bernoulli's principle no longer strictly applies since the viscous friction dissipates energy of the flow, resulting in a pressure drop along the pipe. At the outset, it was thought that this pressure drop would increase in direct proportion to the flow velocity that is, double the flow velocity and the pressure drop doubles as well but experiments showed that this was only true up to a certain threshold. Beyond that threshold, the pressure drop increased with roughly the square of the flow velocity. This hinted that there were two different processes at work, which was demonstrated by Osborne Reynolds, a professor at the University of Manchester in Britain. Osborne devised an apparatus in which he could observe the flow of dye in water pumped through an tube.

He observed that the dye would flow straight for a distance, what was called "laminar" flow, and then break up into a tangle of eddies, what was called "turbulent" flow. This phenomenon can be observed by watching smoke float up off the tip of a lit cigarette: the smoke will rise for a space and then break into a tumble of coils. Incidentally, turbulent fluid flows are an example of a chaotic system.

In any case, the drag of the flow increases dramatically when the flow changes from laminar to turbulent. Reynolds was able to devise an equation that factored in the flow velocity, the density of the fluid, the viscosity of the fluid, and the length of fluid flow to return a value that would indicate when the flow changed from laminar to turbulent. The "Reynolds equation" is given as follows:

Fluid velocity × length of fluid flow × fluid density / fluid viscosity

The value returned by the Reynolds equation is known, to no great surprise, as the "Reynolds number (RN)". If the RN is less than 2,100, the flow will always be laminar. It doesn't necessarily switch to turbulent just above that value, but it becomes increasingly unstable and likely to shift to turbulent as the RN increases beyond 2,100.

One of the interesting features of the RN is that fluid flows with similar RNs behave in much the same way, no matter if the fluid is air or water or whatever. A sailboat moving at low speed in a dense fluid like water has much the same behaviour in terms of fluid dynamics as an aircraft moving much faster in a thin fluid like air. This has a major, though not obvious, implication for issues of scaling in aircraft design.

The smaller an aircraft gets, the greater the effective viscosity of the air. To a gnat, the air feels thick and viscous, and flying is more like swimming in fact, the smaller the creature, the less bother it is to get into the air, and small spiders will sometimes migrate by simply throwing out a thread of silk to the wind. Researchers working on tiny robot aircraft have finding out that conventional aircraft designs are not very workable at such small scale, and they have gravitated towards unusual designs, such as circular flying wings, or even "ornithopters" that flap their wings to fly.

- The complexity of the analysis of viscous flows meant that scientists and engineers really needed to obtain simplified but still workable methods. In 1904, the German engineer Ludwig Prandtl observed that flows could be often be separated into two regions: a thin region known as the "boundary layer" close to the surface over which the flow passes and where the viscous effects are concentrated, and the region of flow above the boundary layer where viscous effects could be disregarded.

It is this viscous boundary layer that causes the airflow over the top of a wing to follow the curve of the wing in effect, it causes the airflow to stick to the wing. At low speeds, the boundary layer will tend to "separate" from the wing and ruin lift, and so a number of combat jets were built with a "blown flaps" or "boundary lsayer control (BLC)" scheme in which jet engine bleed air was blown over the top of the wing or the flaps to ensure that low-speed separation didn't occur.

The boundary layer can also be a nuisance for combat jet aircraft. The engines of such aircraft require a continuous large airflow to provide a high level of thrust, and if the engine intakes are close to the aircraft's fuselage they may swallow the viscous ("stagnant") boundary-layer air, choking off airflow and engine power even leading to "engine stall" that can "flame out" the engine or, in the worst case, damage it. Such aircraft often have a "splitter plate" between the front of each engine intake and the fuselage; these plates are set slightly off from the fuselage and prevent the boundary-layer air from being sucked into the inlet.

- Assuming that a fluid is incompressible is appropriate if the fluid is a liquid, but it can lead to drastically wrong conclusions when the fluid is a gas. Even assuming nonviscous laminar flow, Bernoulli's law only really applies to slow-moving gas flows; if the gas flows are rapid, compression comes into play, leading to distinctly different behaviour.

  Analysis of a compressible flow is too tricky to be discussed in detail in this document, but a simple example will illustrate the principles involved.

Consider a standard rocket nozzle: it can be modeled as a cylinder that leads to a constriction, known as the "throat", which leads into a widening "exhaust bell" open at the end. Hot gases flowing down the hit the throat, where the velocity picks up and the pressure falls, roughly as it would in a fluid, though compression occurs as well.

Once the hot gases exit the throat and enter the widening exhaust bell, they expand rapidly, and this expansion drives the velocity up while the pressure continues to fall exactly the reverse scenario of what would occur if a fluid passed through a constriction into a widening bell.

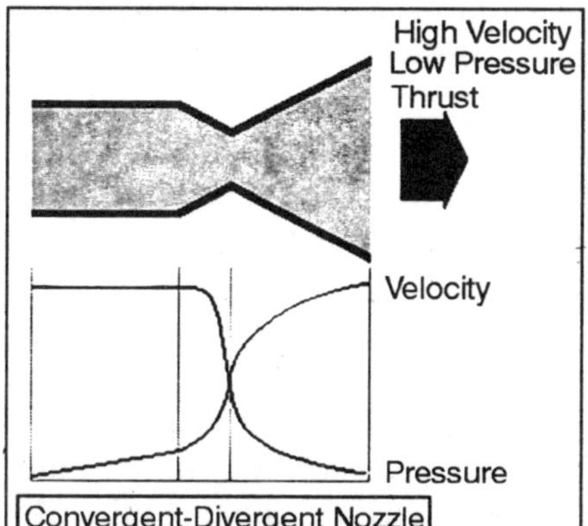

Convergent-Divergent Nozzle

A standard rocket nozzle is designed to provide optimum thrust for a certain range of altitudes and atmospheric pressures.

High-performance jet combat aircraft must operate effectively over a wide range of conditions, and so they have exhausts whose geometry can be varied depending on need.

A fully "authentic" computer simulation of a real-life gaseous flow that takes into consideration viscous effects, turbulence, and compression can require some serious computing horsepower. There is a complicated interaction between all the facets of the simulation which can get even more complicated if chemical transformations take place in the stream.

One aspect to the complexity is that the pressure, volume, and temperature of a gas are closely related to each other. This leads to a consideration of concepts of "temperature" and "heat", which are somewhat devious to define. Although this document has mentioned temperature effects i`n a casual fashion, to go much further requires a more detailed discussion. Similarly, although matter can often be converted between solid, liquid, and gaseous forms through changes in temperature, that also requires a more detailed look.

Chapter 8

Thermal Physics

- One of the simplest, and most difficult to grasp, concepts in elementary physics is that of "heat", and that is the starting point for the study of "thermal physics" or "thermophysics". This chapter provides an introduction to thermal physics.

HEAT / TEMPERATURE SCALES/HEAT ENERGY

- "Heat" and "temperature" may seem to be intuitive concepts, and certainly we do understand them intuitively from daily experience. We can go outside and notice if the temperature is high on a hot day, or low on a cold one. We put food in the microwave or in the oven to heat it up, and we put icecubes in a drink to cool it down. However, properly defining these terms is tricky. Formally speaking, heat is not a property of an object, but is instead a transfer of energy between objects. It follows a few simple rules:
- Heat flows from a hot object to a cooler one. We use a hot plate to boil water, for example, essentially pumping heat into the water.
- Heat never flows in the reverse direction by itself. We have to do work to cool something down, for example using a refrigerator to freeze water into ice cubes.

The great French chemist Antoine Lavoisier postulated that a hot object contained a concentrated amount of a hypothetical "fluid" that he called "caloric" that was the agent of heat transfer. It could be generated by friction, for example rubbing sticks together to start a fire, and could be converted back into mechanical work, for example using a steam engine. The idea seems a bit silly in hindsight, but the notion of "imponderable fluids" was popular at the time, and considering electricity as a fluid, as was the case in that era, was not so ridiculous, since an electric current can be thought of as the flow of a "fluid" of electrons. However, the problem with caloric was that its only identifiable characteristic was that it transferred heat. It was not visible in any way and had no other recognizable properties. What transferred heat? Caloric. What was caloric? It was what transferred heat. Physicists felt they weren't on very solid ground.

It wasn't long after Lavoisier introduced the idea of caloric that scientists began to understand that heat was simply the consequence of the motion of the molecules in an object. Benjamin Thompson was a Loyalist who backed the wrong side in the American Revolution and had to flee to Europe, where he acquired the title of "Count Rumford". Thompson was working in the arsenal at Munich, and noticed that boring out a cannon generated a good deal of heat.

That wasn't any new observation, but he observed that the rate of heating didn't depend at all on the rate at which the cannon bore was ground out, only on how much work was put into the grinding process. A dull tool that ground the bore slowly generated about as much heat at any one time as a sharp tool that ground quickly, if both tools were rotated as fast. From the point of view of caloric theory, the sharp tool that ground quickly should extract more caloric and should release more heat. That wasn't the case, and Thompson realized that he could make the tool duller and duller until he could get heat out of the process indefinitely. That would imply that the cannon blank stored a infinite supply of caloric.

Thompson deduced that the work of the grinding process was being converted into heat, and that caloric was just a fantasy. It would take time for his ideas to be widely accepted, since his writings on the subject were general in nature and not backed up by detailed studies. Others would take up his ideas and put them on a stronger foundation. James Prescott Joule conducted a series of elegant experiments to demonstrate the link between work and heat, culminating in an 1845 paper in which he described an apparatus in which weights were connected to a cord and a pulley arrangement, to spin a set of paddles in a closed vessel full of water. Using a sensitive thermometer to measure the heat produced by the paddles showed it was predictably proportional to the energy released in the fall of the weight.

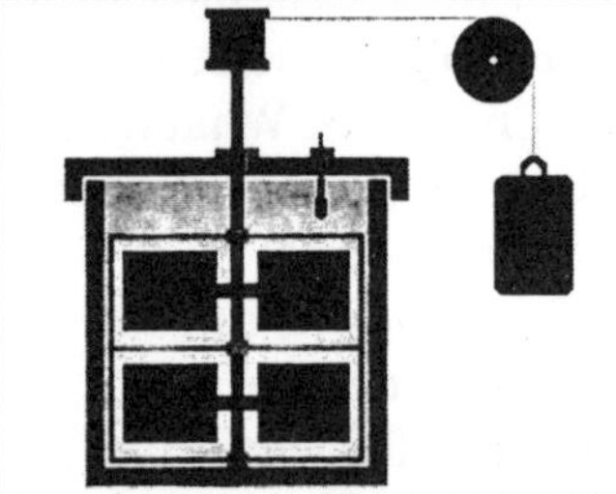

Joule' Experimental Apparatus

It is now known that force generated by friction over time exerts work that produces heat energy, in a cannon or anything else that is affected by friction forces. In more modern terms, adding energy to an object by friction or another heating process increases the velocity and the kinetic energy of the molecules, increasing the "internal energy" or, more informally, the "thermal energy" of the object. The temperature of the object was then a measure of the thermal energy, essentially an average of the kinetic energy of all the molecules of that object.

Heat in turn became a transfer of this thermal energy. If a hot object was brought in contact with a cooler object, collisions between the energetic molecules of the hot object increased the velocity of the slower molecules of the cooler object. One of the interesting implications of the view of temperature as a measure of molecular motion was that if all molecular motion in an object ceased, then there would be no way for the object to get any colder. This implied the existence of an "absolute zero" temperature.

- There are several different temperature scales in use today. The "Celsius" scale, once known as the "Centigrade" scale, is in use over most of the world, while the "Fahrenheit" scale is used in the United States. The "Kelvin" scale, which is closely related to the Celsius scale, is in common use for scientific purposes.

In the Celsius scale, the freezing point of water is specified as 0 degrees Celsius, while the boiling point is specified as 100 degrees Celsius. In the Fahrenheit scale, the freezing point of water is 32 degrees Fahrenheit and the boiling point is 212 degrees Fahrenheit. Conversions between the two can be performed as follows:

degrees Fahrenheit = (9/5)×degrees Celsius + 32

degrees Celsius = (5/9) • (degrees Fahrenheit–32)

The Kelvin scale is the same as the Celsius scale, except that the 0 point is at absolute zero, which is equivalent to -273.15 degrees Celsius. This makes conversion between the Kelvin and Celsius scales simple:

degrees Kelvin = degrees Celsius + 273.15

degrees Celsius = degrees Kelvin – 273.15

The Celsius and Fahrenheit scales are useful for describing temperatures in our daily environment, since they give small values for most Earthly temperature ranges.

The Kelvin scale is much more convenient for scientific work, since it eliminates negative temperature values that are clumsy in calculations. Incidentally, there is also a "Rankine" scale, now almost never seen, that is the same as the Fahrenheit scale, except that the 0 point is at absolute zero, equivalent to -459.67 degrees Fahrenheit.

- In metric units, heat and thermal energy are given in terms of the "calorie", which is the amount of heat required to raise a gram of water from 15 degrees Celsius to 16 degrees Celsius at a pressure of one atmosphere. Since heat is equivalent to energy, one calorie is also the same as 4.186 joules. Since the calorie is a somewhat small unit for most practical purposes, it is often expressed in terms of "kilocalories", and just to confuse matters, the "Calorie" used to rate food energy actually means a kilocalorie.

In English units, heat is given by the "British thermal unit (BTU)", which is the amount of heat required to raise a pound of water one degree Fahrenheit. The BTU has been generally obsoleted by the calorie. One BTU is equivalent to 252 calories or 1,055 joules.

HEAT AND MATTER

- Heat of course has a number of effects on matter, most visibly on gases. An "ideal gas", meaning one where the molecules that are visualized as perfectly elastic particles that have no tendency to attract each other, has a neat relationships between temperature, volume, and pressure. This can be summarized as the "ideal gas law":

$$\frac{\text{pressure1} \times \text{volume1}}{\text{temperature1}} = \frac{\text{pressure 2} \times \text{volume 2}}{\text{temperature 2}}$$

For example, suppose Dexter stores a gas in a rigid airtight container with solid walls meaning its volume is constant and that the temperature of this gas is the same as the ambient temperature of the outside world. If Dexter then heats up the gas inside the container to twice its original temperature (using the absolute Kelvin scale), them the final pressure is twice the original pressure. Similarly, if Dexter cools the container to half its original temperature, the pressure falls to half of its original pressure. This is a "constant volume" process:

Pressure 1 / temperature 1 = pressure2 / temperature2

One interesting example of this process at work is to pipe hot steam into a can and then seal it. If the can's walls are thin, once the steam cools off, the pressure in the can will drop, and the external atmospheric pressure will crush the walls of the can.

Now suppose the top of Dexter's rigid container is really an airtight piston that can slide up or down in the container. Dexter again heats up the gas inside the container to twice its original temperature, but this time he lets the piston out so the pressure remains constant as the heat increases. Assuming that he does this quickly so that none of the heat leaks out, the end volume is twice that of the original volume.

Similarly, if Dexter cools the gas and pushes in the piston to maintain a constant pressure, assuming no heat leaks in, the end volume is half the original volume. This is a "constant pressure" process:

volume 1 / temperature 1 = volume 2 / temperature 2

Now suppose Dexter pulls out the piston to expand the container to twice volume and then lets the gas inside warm up to the outside ambient temperature. The end result will be that the pressure falls by half. Similarly, if he pushes in the piston to compress the container to half volume and lets the gas cool off to the outside ambient temperature, the pressure doubles. This is a "constant temperature" process:

Pressure 1 Xvolume 1 = pressure 2 Xvolume 2

The relationship between pressure and volume, temperature being constant, was observed by the English chemist Robert Boyle, who suggested it in a book published in 1661. The volume and temperature relationship, pressure being constant, was suggested by the French chemist Jacques Charles in 1787, and the pressure and temperature relationship, volume being constant, was proposed by another French chemist, Joseph Gay-Lussac, who published it early in the 19th century and came up with the combined gas law to cover all three laws.

The relationships in the combined gas law become more complicated if temperature, volume, and pressure are varied all at the same time this is one of the reasons why calculating the effects of compressible flows is so difficult but the general effects of changes in pressure, temperature, and volume remain apparent. It should be remembered that gases at typical Earthly conditions are a good approximation of ideal gases, but at extremes of pressure or temperature gases may depart from this nice neat behaviour.

- The ideal gas law was derived empirically, determined from experiments not too much different conceptually from the experiments Dexter performed above. The idea that heat was due to molecular motion did obviously imply a link between the ideal gas law and classical Newtonian mechanics.

  Newton's principia mathematica pointed out that the pressure-volume relationship discovered by his contemporary Robert Boyle could be explained if a gas was considered as a collection of tiny elastic particles, flying around in free space. However, making a rigorous connection between the two was tricky.

It was possible to model the behaviour of a single molecule using Newtonian mechanical principles, envisioning it as a particle with a given velocity that undergoes collisions. The problem was that a gas is made up of a vast number of particles that all have different speeds. Trying to measure or model the individual behaviour of each molecule in a gas was obviously impossible.

The solution to this problem was discovered by three 19th-century physicists: James Clerk Maxwell, a Scotsman ("Clerk" is pronounced "Clark" in this case, incidentally); Ludwig Boltzman, an Austrian; and Josiah Willard Gibbs (1839;1903), one of the first major American theoretical physicists.

Their approach was to perform a statistical analysis on the large number of particles, using the basic laws of Newtonian mechanics to obtain a "probability distribution" that gave the relative proportions of particles in a given range of velocities for a given temperature. Although tracking all the particles was impossible, it was possible to statistically determine their average behaviour.

The analysis, now known as "Maxwell-Boltzman statistics", allowed the ideal gas law to be derived from Newtonian mechanics, linking the "microscopic" world of particle interactions with the "macroscopic" world of the ideal gas law. The details of Maxwell-Boltzman statistics are not very relevant to an introductory physics text, since the ideal gas law is all a casual student of physics really needs to know.

The idea is relevant in the broader sense that it introduced the concepts of statistical analysis to fundamental physics: the notion that, although there might be no way to determine the action of the specific particles of a system, the "law of large numbers" would still allow the behaviour of that system to be predicted. This powerful technique would be exploited in new directions by 20th-century physics.

- Liquids and solids are generally not very compressible, and so do not follow the same laws as gases. Heat still has an effect on them. Most solids and liquids increase in volume when heated, and decrease in volume when cooled. This is why bridges have "expansion joints" at intervals, to compensate for the changing length of the span due to temperature changes. The change in the size of an object made of a particular material is called its "coefficient of thermal expansion".

One simple practical application of this property is the "bimetallic coil" used in some thermometers. This is a coiled strip of metal with brass on one side and iron on the other. Since the two metals have different coefficients of thermal expansion, as the strip is heated or cooled it will bend one way or another in a predictable fashion, and the raveling or unraveling of the coil could be used to turn a thermometer needle.

Bimetallic Coil Thermometer (GVG/PD)

The same principle is used in traditional thermostats, though they use an uncoiled bimetallic strip. The thermostat can be set so that if the strip bends to a particular position, it will complete an electric connection and turn on the heat or air conditioning as need be.

- Heat of course can change a substance from solid to liquid and from liquid to gas. For example, water is solid ice at low temperatures, but changes to liquid at higher temperatures, and steam at even higher temperatures. Different materials go through such "changes of state" or "phase changes" at specific temperatures.

  Some materials don't go through a liquid phase, converting directly from solid to vapour, at least under typical Earthly atmospheric pressures, a process that is known as "sublimation". Frozen carbon dioxide sublimates, which is why it is used to cold-pack parcels that have to be shipped, since such "dry ice" doesn't create puddles. Incidentally, water ice sublimates at a low rate: leave an ice-cube tray in a freezer for many months, the ice cubes will gradually fade away, or at least they will if the freezer is opened often and the climate is dry.

The amount of heat that must be added to a substance produce a phase change is called the "latent heat". Different phase changes have different latent heats: a particular material will have different latent heats of melting, vaporization, or when it applies, sublimation.

At standard atmospheric pressure, water cannot be heated above 100 degrees Celsius, because any additional energy simply vaporizes the water. Boiling water is actually, in a sense, a "cooling" process since it prevents the accumulation of energy. A pressure cooker is used to obtain higher water temperatures. When water condenses again, the latent heat that vaporized the water is released.

The proportion of water vapour in the air is referred to as "humidity". There is a maximum level of humidity that air can support, which increases with temperature since water condenses back to liquid more easily at low temperatures than high. When the air can accommodate no more water vapour, it is said to be "saturated". Humidity is usually given in terms of "relative humidity", or the ratio of actual water vapour to the saturation level. At 50 per cent relative humidity, the air contains half the amount of water vapour that it is capable of supporting, and at 100 per cent relative humidity the air is saturated.

The evaporation of water is used for cooling in the "evaporative cooler" or "swamp cooler", a simple household cooling device used in dry, warm climates. It consists of little more than a fan pulling air through a filter through which water is pumped.

The air causes the water to evaporate, moistening and cooling the air. It is substantially cheaper to buy and operate than a conventional air conditioner, whose operating principles will be discussed in the next chapter. An evaporative cooler is ineffective in hot humid climates since the rate of evaporation slows, and in fact it may simply make a living space more humid and sticky.

Incidentally, water is an unusual substance in that it actually expands when frozen, due to the way its molecules rearrange themselves. This is a fortunate circumstance for life on Earth, since if it were not so, all ice forming on the ocean would sink to the bottom and gradually build up a reservoir of ice that would make the planet a permanent "icebox".

- Gases, liquids, and solids that are being heated, but not changing state, have a "specific heat" that defines how much energy must be added per unit mass to raise the temperature of unit mass of the substance one degree. The specific heat of water, for example, is by definition one calorie per gram per degree Celsius. Incidentally, the specific heat of water is unusually high, and so it takes more energy to heat water than most other substances. In the case of materials that are compressible, specific heat has to be rated in terms of whether the materials are held at constant pressure or at constant volume.

The concept of specific heat leads to the notions of "thermal mass" and "heat reservoir". For example, a lake has a thermal mass in that it will require energy and involve a certain delay to heat it up, and once heated up will act a reservoir of heat, slowly releasing it back to the environment.

HEAT TRANSFER

- Heat is transferred by three processes:
- The first is simple "conduction", where it is transferred through direct contact.
- The second is "convection", in which currents are set up in a material that transfer heat through it, like the circulating currents in a pot of soup.
- The third is "radiation", in which heat is transferred through space by an invisible form of light known as "infrared radiation". This is the heat projected by a heat lamp or a glowing electric heater. The hotter the object, the more radiation it emits, and the more the radiation creeps up from the invisible infrared radiation into the visible range. Infrared radiation is discussed in more detail later.

Solid and opaque objects support heat transfer primarily through conduction. Different materials have specific "thermal conductivities", or rates of heat flow through them. For example, metals generally have high thermal conductivities: if Dexter wears metal-rimmed sunglasses on a cold day, they will painfully drain the heat right out of his nose. Plastics generally have lower thermal conductivities, and it is much more comfortable to wear plastic-rimmed sunglasses on a cold day.

Materials with low thermal conductivities are, naturally, used as thermal insulators for homes and buildings. Gases generally have low thermal conductivities, and most forms of home insulation are basically designed as "gas traps". The most vivid example of this principle is sheet plastic foam insulation, which is filled with bubbles of gas.

Incidentally, the insulating capability of commercial insulation is specified by an "R-value", which is inversely related to the thermal conductivity. The higher the R-value, the better the insulator. For example, thick fiberglass insulation has an R-value of almost 20, while double-paned glass has an R-value of a little over 1. For a high degree of thermal insulation, a "dewar flask" is used.

The common consumer "Thermos bottle" is a cheap type of dewar flask. A lab-quality dewar flask consists of a double-walled vessel made of Pyrex glass with silvered surfaces and the space between the two walls evacuated. A vacuum has *no* thermal conductivity; of course it can't support convection; and the silvered surfaces reflect radiation and so limit loss by that route.

The neck and cap of the flask end up providing most of the thermal conduction, and so are generally made as small as possible.

Extreme cooling requires a "double dewar" scheme, with one Dewar flask contained in a second, and the space between the two filled with a cryogenic fluid such as liquid nitrogen (with a boiling point of 77 degrees Kelvin) or, for really cold applications, liquid helium (with a boiling point of 4 degrees Kelvin).

Infrared telescopes launched into orbit around the Earth require extreme sensitivity to observe distant and cool cosmic objects. One of the problems with infrared telescopes and other infrared imagers is that they can't pick up a target that's cooler than they are, since their own thermal emission will drown out the infrared emitted by the target.

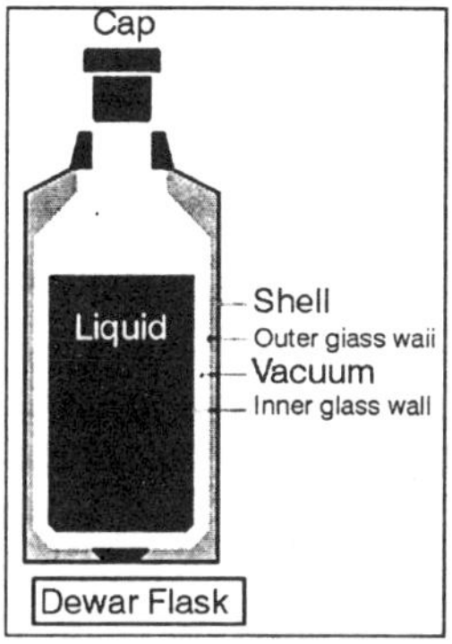

As a result, space infrared telescopes are often built as (very large) double-dewar flasks known as "cryostats", with an infrared telescope built inside the inner dewar flask. They may also have a reflector shield made of multiple layers of plastic, separated by spaces exposed to the vacuum, to reduce solar heating and conserve their cooling helium.

As mentioned in an earlier chapter, they are often put into halo orbits at the Earth-Sun L2 point so they can always keep the shield pointed at the Sun.

- The physical configuration of an object affects its ability to retain heat. The simple general rule is that an object retains heat better if the ratio of its surface area to its volume is kept as low as possible. This is why people adapted to cold climates tend to be short and stout, while those adapted to hot climates tend to be tall and slim, giving them a greater ratio of surface area to volume to help them get rid of heat through convection and radiation.

Air-cooled engines will have sets of fins around the cylinders to allow the heat to be dissipated. Similarly, inspection of the circuit board for, say, a personal computer, will reveal that the processor chip is usually fitted with a finned plate or "heatsink" to dissipate its heat of operation. A personal computer will also generally have an electric fan to get rid of the warmed air inside of the enclosure and draw in cooled air.

THE LAWS OF THERMODYNAMICS / ENTROPY AND THE ARROW OF TIME

- The behaviour of heat in a physical system is described by three rules, known as the "Laws of Thermodynamics". The Laws of Thermodynamics govern the efficiency of engines, and also rule out "perpetual motion machines".

Thermodynamics is an abstract field even by the standards of physics. To understand it, a few formal definitions must be set down first. The most important is that of a "thermodynamic system", which is defined as a domain bounded in space where heat can flow across the boundaries in either direction.

The properties of a thermodynamic system at any one time define its "state". The most important properties are the "thermodynamic variables" of temperature, pressure, and volume, but there are other variables, such as density, specific heat, and the coefficient of thermal expansion. If the properties of a thermodynamic system do not change over time, and if there are no changes in its configuration and no net transfers of heat across the boundary of the system, the system is said to be in "thermal equilibrium". If the thermodynamic system moves from one state of thermal equilibrium to another, a "thermodynamic process" is said to have taken place. A thermodynamic process can be "reversible" or "irreversible". A reversible process can be run in one direction, and then be reversed to return to the same state that it started from with no net change in system energy. An irreversible process can be run in one direction, but will require a net input of system energy to reverse it.

- There are four Laws of Thermodynamics. Originally, there were only three, but later physicists decided that a fourth law was required. Since this law was basic to the other three, it was called the "Zeroth Law of Thermodynamics" instead of the "Fourth Law of Thermodynamics". The Zeroth Law is basically a definition of the term "temperature". It states that if two thermodynamic systems are in thermal equilibrium with a third thermodynamic system, then the first two systems are in thermal equilibrium with each other. They will all share the same "temperature".

- That formality out of the way, the "First Law of Thermodynamics" defines heat. The First Law of Thermodynamics stems from the work of Joule, Mayer, and Helmholtz, and amounts simply an extension of the law of conservation of energy to include heat.
  It states that when a warm object (thermodynamic system) is brought into "thermal contact" with a cooler object, meaning a connection that allows the transfer of heat, a thermodynamic process will take place that eventually brings them to the same temperature. The process transfers "heat energy" between the two objects.

The amount of heat transferred into a system, plus the amount of work done on the system, must result in a corresponding increase in the thermal energy of the system. The First Law further implies that if any work is done by the system, it must drain the internal energy of the system. A system where work can be done without draining the internal energy of the system is referred to as a "perpetual motion machine of the first kind". The First Law rules out such machines.

- The "Second Law Of Thermodynamics", established through the work of the English physicist William Thompson, later Lord Kelvin, for whom the "Kelvin" temperature scale is named, and the German physicist Rudolf Julius Emanuel Clausius. It is by far the most important of the four laws, and some have suggested that it really *should* have been the First law, not the Second, though in practice the Zeroth and First laws act as necessary steps up to it.

The Second Law requires definition of a property known as "entropy" that was devised by Clausius. In formal thermodynamic terms, it is defined as:

$$\text{Entropy} = \frac{\text{heat\_transfer}}{\text{absolute\_temperature}}$$

In much more informal terms, entropy measures the thermodynamic "disorder" of a system, or roughly speaking the tendency of the system's energy to degrade into random thermal energy. The greater the entropy, the greater the disorder.

The significance of the specific form of this definition will be discussed in the next chapter. Entropy can actually be expressed in other ways, and the definition of the term has to be carefully considered for any given scenario.

The Second Law states that the entropy of a "isolated" or "closed" system, meaning one where there is no transfer of energy across its boundaries, can never decrease. It may remain the same, or it may increase. A closed system where the entropy decreases is referred to as a "perpetual motion machine of the second kind". The Second Law rules out such machines even though they do not violate energy conservation more on this in the next chapter.

- The "Third Law of Thermodynamics" is a bit of an anticlimax after the other laws. It simply recognizes the existence of the absolute temperature scale, and states that absolute zero can never be attained in practice in any finite number of cooling steps. A cooling system can in principle approach absolute zero by a narrower and narrower margin, but it will never actually reach absolute zero.
- In modern times, as mentioned earlier, the concept of caloric has been replaced by an understanding of heat as the motion of the molecules of a system, using statistical methods based on classical mechanics to determine the average behaviour of a large number of molecules. In this view, temperature is a measure of the average kinetic energy, or essentially the motion, of the molecules of a system. A temperature increase means that the average kinetic energy has increased.

Similarly, heat transfer between two thermodynamic systems is caused by collisions between individual molecules of the systems. The collisions continue until on the average the net energy passing across the boundary between the two systems is zero, meaning the two systems have achieved thermodynamic equilibrium.

The First Law of Thermodynamics, then, exactly corresponds to the classical law of conservation of energy. The molecular nature of the Second Law of Thermodynamics is a little more subtle, and basically expresses entropy as the measure of the "probability" of a system. Given a large number of molecules in a thermodynamic system, probability dictates that the molecules will be moving in many different directions (high entropy) than in one direction (low entropy).

- The Second Law sets a direction of time, or "arrow of time", in the direction of increasing entropy. On the microscopic atomic scale, interactions are generally reversible: if we make a video of a collision between two atoms, it will be just as valid if the video is run forward or in reverse. At a macroscopic level, this can be also seen in, say, collisions of billiard balls: if a video is made of two balls colliding, ignoring friction it will be valid if run forward or in reverse.

However, the disorder implicit in entropy ensures that in our macroscopic world time only runs in the forward direction. If we make a video of a sugar cube dissolving, it is completely obvious whether the video is being run forward or in reverse. The same is true of the "break" at the beginning of a billiards game, in which the orderly balls are scattered over the table.

Modern physics would show that this arrow of time had a catch, and also show the solution. Given a system with a low entropy, or low probability, in the present, then it is of course likely to evolve to a state with high entropy, or high probability, in the future.

ENTROPY & SUGAR CUBES (GVG / PD)

This is what the Second Law basically says. The catch is that, given a system with low entropy / low probability in the present, then the odds would suggest that the system also had high entropy / high probability in the past. Of course, this isn't the way things really work, as a rule, in the past the system had even lower entropy and probability.

It was the discovery of the Big Bang, the primordial explosion that created the Universe we live in, that provided the explanation. The Universe began in a state of high energy and uniformity with low entropy, or in other words the Big Bang provided a "zero reference" for entropy, and the entropy has been increasing ever since.

Put simply, the Universe is running thermodynamically downhill from an initial height, and that defines the arrow of time on the macroscale. Some have speculated that if the Universe were to cease to expand and fall back in on itself, then time would reverse.

This is getting well off track. This whole notion of the arrow of time has been considered at length and in great subtlety by modern physics, but in classical physics it is taken as a given that time only flows in the forward direction. From the microscopic physics point of view, there is no reason that the motion of all the molecules in a brick that has been dropped to the ground could not point straight up, causing the brick to fly back spontaneously into the air. However, this configuration is vanishingly improbable. It simply doesn't happen it is an irreversible process.

Similarly, there is no reason on the microscopic scale that air molecules could not transfer net heat into a brick, making it warmer, but this is improbable as well. A hot brick will release its heat into cooler air, but cooler air will not warm up a brick. Heat always transfers from a warm object to a cooler one, never the reverse. In fact, this is actually just another way to state the Second Law.

- The probabilistic nature of entropy can be illustrated in a simple mathematical fashion with a system consisting of two containers linked by a tube. One of the containers contains four balls colured red (R), green (G), blue (B), and white (W).

The tube is big enough to allow the balls to pass back and forth between the containers, and if the system is shaken for an extended period of time the balls will be distributed between the two containers in a random way.

For the four colurs of balls, there are two results in which all four balls are all either in one container or the other, as represented below:

[R G B W] [– – – –] [– – – –] [R G B W]

There are 8 results in which there are three balls in one container and one ball in the other:

[– G B W] [R– – –] [R– – –] [– G B W]
[R–B W] [– G – –] [– G – –] [R – B W]
[R G – W] [– – B –] [– – B –] [R G – W]
[R G B –] [– – – W] [– – – W] [R G B –]

There are 6 results in which there are two balls in one container and two balls in the other:

[– – B W] [R G – –] [R G– –] [– – B W]
[– G – W] [R – B –] [R – B –] [– G – W]
[– G B –] [R – – W] [R – – W] [– G B –]

There are a total of 2 + 8 + 6 = 16 possible results, which gives the probabilities:

Four balls in one container: 2/16 = 0.13

Three balls in one container, one in the other: 8/16 = 0.5

Two balls in each container: 6/16 = 0.38

As the number of balls in this system increases, the probability of ending up with all the balls in the same container approaches zero. With all the balls in one container, the system has low entropy; with the balls divided between the two containers, it has high entropy.

As Boltzman pointed out, such a probabilistic approach can also to be used to construct an alternate definition of entropy:

entropy = constant × natural log (number of available states)

In the "macroscale", Boltzman's definition of energy gives effectively the same results as Clausius's definition. (There is also a third definition for entropy from a branch of theoretical math known as "information theory", but this definition can be seen as a generalization of the Boltzman definition to cover phenomena outside of thermodynamics, and it doesn't bring anything new to the party for thermodynamics as such.)

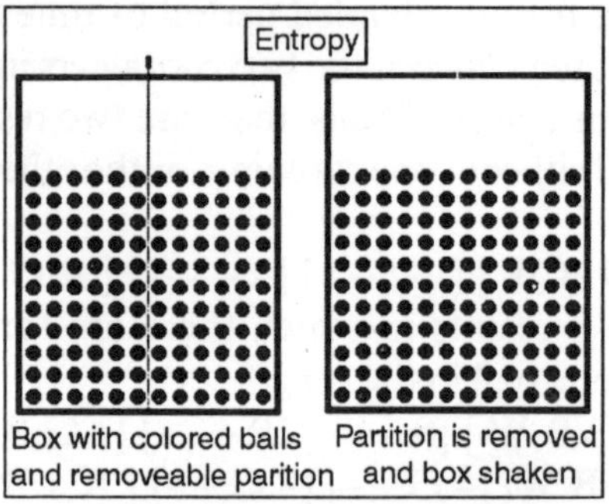

Like the Maxwell-Boltzman statistics, the Boltzman definition of entropy is of little use for simple physics: it is simply given here as a connection to more detailed studies. No matter how entropy is expressed, the essential point is that it is easy to dissolve, mix, or disperse materials, but it is hard to sort them out or bring them together again.

FOOTNOTE: ENTROPY AND EVOLUTION

- The Second Law of thermodynamics is occasionally cited as a disproof of Charles Darwin's theory of evolution, in that evolution seems to imply order rising spontaneously out of disorder.

Actually, the term "evolution" is a bit ambiguous, since by itself all the word means is that there has been a progression of species from common ancestors on the Earth, demonstrating a broad increase in diversity and elaboration. Evolution is about as much a scientific fact as anything in the sciences, since the fossil record shows a clear succession of forms, and the various structural relationships between different life-forms don't make any visible sense without assuming "common descent". The question is *how* it happened:

Darwin proposed that evolution has been driven by spontaneous mutations in new generations of organisms, with the least successful dying out and the most successful surviving to create a next generation. There's plenty of evidence to suggest that "Darwinism", as evolution by natural selection is sometimes called, is not an unreasonable idea. Certainly human selective breeding has been able to transform creatures all out of recognition, pekinese out of wolves as a vivid example, by doing no more than exploiting natural mutations. Darwin never claimed any more than that natural forces the competition for food, tolerance to climate conditions, the ability to escape predators, the ability to find a mate would perform a similar selection.

None of the individual actions, including reproduction, mutation, and selection by natural forces, violate the Second Law in fact, all are easily observed and in themselves entirely provable. If none of the components of the process violate the Second Law, is there any way to show that they do so in combination?

In practice, evolution by natural selection is clearly demonstrated by the spontaneous emergence of antibiotic-resistant bacteria, and of bacteria that can digest synthetic materials such as nylon. Darwinism admittedly does claim that more elaborate life-forms are generated from less elaborate ones by "trial and error" processes. Critics find this idea implausible, and while most are willing to admit that it accounts for the emergence of antibiotic-resistant and nylon-digesting bacteria "microevolution" they claim that it cannot account for the full elaboration of evolution "macroevolution". However, if the Second Law ruled out macroevolution, it's hard to see how microevolution could still work: what the Second Law shows to be absolutely impossible on the large scale should also be absolutely impossible on the small. Even dismissing microevolution, it's hard to see how the Second Law could rule out the emergence of elaborate multicellular animals from single-celled ancestors since by the same logic it would similarly rule out the development of a complete human being from a single-celled egg.

The two aren't the same thing, of course, but feature the same spontaneous progression from simple to elaborate. If one is a violation of the Second Law on the face of it, then the other should be as well. Critics claim that it is the "intelligence" of the genetic organization of a single-celled egg that allows it to grow up into a complete human; evolutionists reply that it is the "intelligence" of natural selection that shapes a "signal" out of random mutations into new adaptations of organisms.

In fact, it is very hard to identify any violation of the Second Law by Darwinian evolution on the face of it. The Second Law specifies that entropy for a closed system must increase over time.

Entropy, in the classic definition, is strictly a thermal concept: heat transfer divided by absolute temperature. There is no thermodynamic reason that biological systems may not acquire more elaborate levels of organization over time; all the Second Law says is that they will waste a lot of energy doing it, in the same way a baby eats a lot of food to grow up into an adult.

The reality is that the metabolism, growth, and reproduction of organisms, or in other words the basic mechanisms of Darwinian evolutionary processes, are all busily increasing entropy through their consumption of energy and there is a line of thermodynamic thought that suggests that thermodynamic "gradients", between low and high entropy, often set up complicated phenomena in the first place.

In this view, life a particularly complicated phenomenon is simply a "trap" between the low-entropy energy provided by sunlight, which drives the food chain through plant photosynthesis, and its ultimate degradation as waste environmental heat. These are all informal arguments, but that observation cuts both ways: nobody has ever produced a formal mathematical analysis to show that the evolution of life violates the Second Law that has been able to stand up to criticism. Nobody been able to come up with a particularly meaningful way to put a value on the entropy of a biological system to permit a valid thermodynamic analysis; nobody's been able to come up with any particularly useful scheme to quantify in any way the complexity of biological systems, or in other words assign a useful value of the complexity of an amoeba versus a human being. From a genetic and structural point of view, a mouse is about as complicated as a human being: a human is bigger, smarter, and lives far longer, but the genetic codes of the two species are roughly the same size, they feature all the same basic systems, and at a functional level the two organisms operate in much the same way. Whether Darwinism is valid or not, the Second Law neither disproves nor proves it.

- There is, however, another question involved in this matter, the issue of how life got started in the first place.

> This is not an issue with Darwinism as such, since Darwin never claimed he had any idea of how life arose. His theory of the mechanisms of evolution simply assumed that life arose in some unspecified fashion and he only attempted to describe what happened after that in much the same way that a book on computer programming doesn't explain how to build a computer.
>
> Darwin thought it possible that the origin of life would always remain a mystery. It still is; researchers have a number of speculative theories, but none are regarded as complete or fully convincing.

The basic idea is that life arose under special conditions that encouraged the creation of ever more elaborate systems of molecules with this spontaneous emergence of life from nonlife referred to as "abiogenesis". The Second Law does not inherently rule this idea out.

The initial creation of life from the organization of simple molecules into more complicated ones does imply a decrease in entropy, but only for the molecules themselves, and not necessarily for the complete thermodynamic system in which they exist. Although physical systems do tend in general towards greater disorder, it is still not impossible for order to arise spontaneously from disorder under specific circumstances. Consider the example of a mix of red and white balls given above. Obviously, it can shaken forever and the balls will not resort themselves back into red and white except if the red balls are twice as big as the white balls, in which shaking the mix will cause the red balls to rise to the top and the white balls to fall to the bottom. (Why this happens is actually an interesting question in itself, but irrelevant to the current argument and not discussed here.) The mix has resolved itself into order, but the Second Law is not violated, since the process required an input of energy that resulted in a net increase in entropy.

Orderly crystals can easily be created out of dispersed solutions of atoms, and of course water molecules in the air can freeze into very orderly and often elaborate snowflakes; nobody sees such things as miraculous. If the atoms have a selective attraction to each other, they will tend to join together in a crystal, and a thermodynamic analysis will show that entropy increases in the overall system the snowflake releases heat during its formation, producing an overall increase in entropy.

A snowflake is referred to as an "emergent system", one that exhibits self-organization. Simple lab experiments have been performed in which complicated molecules have been spontaneously self-assembled out of simple molecules through electric sparks. The Second Law is not violated because energy inputs are used to promote the reactions that created the molecules. It is important to realise, however, that such examples do not constitute the slightest proof of abiogenesis: comparing the self-organization of snowflakes to the architecture of a living system is like comparing the complexity of a set of wind chimes to that of an elaborate music-box indeed, such a comparison is a gross understatement.

The origins of life remain mysterious, and it is not unreasonable to retain a little healthy skepticism that the self-assembly of molecules, even over billions of years, could produce living organisms, However, since nobody really knows what happened, it's equally impossible to say that the Second Law or for that matter any other law of physics either proves or disproves it.

Chapter 9

Heat Engines and Refrigeration Systems

This chapter follows up the discussion of the basic laws of thermal physics by considering heat engines, both theoretical and practical, and refrigeration systems.

HEAT ENGINES AND THE SECOND LAW

- All mechanical engines are "heat engines", in which heat is converted to work. Automotive engines burn fuel to set the vehicle rolling, while jet engines burn fuel to generate thrust to keep the aircraft flying. Such engines obtain heat from a high-temperature reservoir, derive work from it, and then pass the waste heat on to a low-temperature reservoir. The Second Law of Thermodynamics has particular application in the analysis of the operation of heat engines.

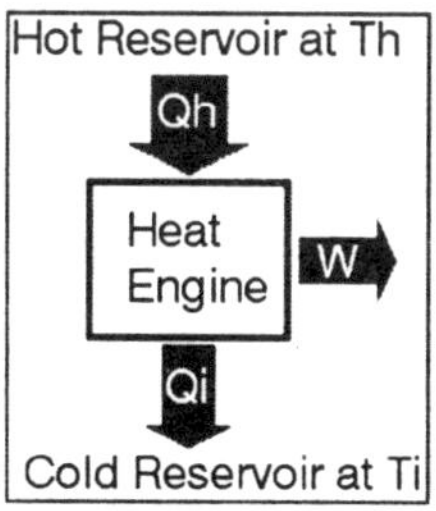

In an "ideal" heat engine, all the heat is converted to work. However, the French engineer Nicolas Leonard Sadi Carnot demonstrated that there is no way to build a 100 per cent efficient heat engine. In a work published in 1824, Carnot showed that the efficiency of an engine is proportional to the temperature difference between the input and output, and to obtain 100 per cent efficiency the difference would have to be infinite.

For example, a power turbine uses steam obtained from a boiler heated by coal or oil to drive the turbine, with the exhaust of the turbine consisting of steam that has cooled in the process. Ideally, the heat flowing out of the process is equivalent to the heat flowing into the process minus the work done:

Heat out = heat in–work

Stated another way:

Work = heat in–heat out

The efficiency of the engine is given by the proportion of the heat input that is actually converted to work, or:

$$\text{Efficiency} = \frac{\text{work}}{\text{heat\_in}} = \frac{\text{heat\_in} - \text{heat\_out}}{\text{heat\_in}} = 1 - \frac{\text{heat\_out}}{\text{heat\_in}}$$

Obviously, the efficiency is maximum when the heat output is zero meaning the heat input is completely used up but this never happens in practice, and this circumstance can be theoretically shown to be impossible. This is a consequence of the Second Law, though the analysis is too complicated to be presented in this document.

- Carnot established these basic rules by performing a theoretical analysis of a simple "ideal" heat engine that provides a basis for the discussion of all heat engines. The "Carnot engine", as it is now known, is a simple cylinder plugged by a piston and containing a quantity of gas, or "working fluid".

Heat engines operate on "thermodynamic cycles", or steps of thermodynamic processes that operate in a circular fashion. The Carnot engine cycles through four phases of operation, as shown at the left side of the illustration below. The changes in pressure and volume for each phase are shown in the "PV (pressure-volume) chart", also known as an "indicator diagram", at the right side of the illustration. In more detail, the cycle works as follows:

- Input of heat "Qh" at high temperature "Th" results in constant-temperature (isothermal) expansion, raising the piston.

- Heat input is removed, and gas continues to expand, raising the piston further. Heat does not escape from the system during this phase of the cycle, or in other words this phase is an "adiabatic" process.
- Loss of heat "Ql" at a low temperature "Tl" causes the gas to compress in an isothermal fashion.
- Heat no longer flows out of the cylinder but the gas continues to compress as an adiabatic process. The engine then starts over at phase 1.

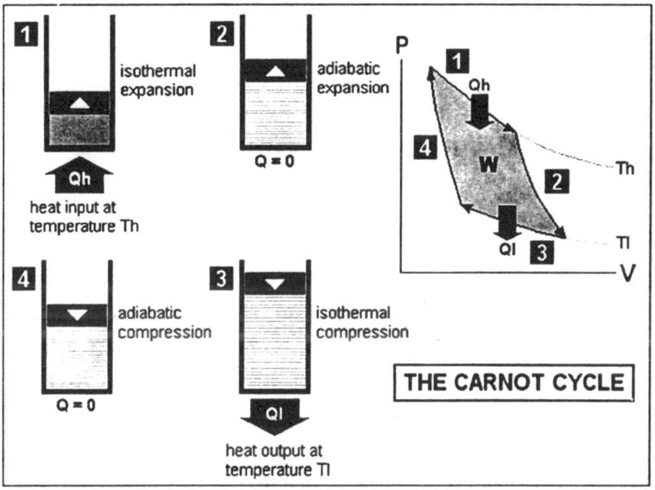

The amount of work W performed by the Carnot engine is given by the area enclosed by the PV curve. This work is equivalent to the input heat Qh less the output heat Ql. As above, this means the efficiency of the Carnot engine is:

$$\text{efficiency} = 1 - (Ql / Qh)$$

Carnot's mathematical analysis of his engine showed that the ratio of input and output heat of his engine was the same as the ratio of input and output temperatures, and so:

$$\text{efficiency} = 1 - (Tl / Th)$$

What this means is that the Carnot engine becomes more efficient as the difference between the input and output temperatures increases. It also shows that an engine can never be 100 per cent efficient, since that would imply a Tl of zero degrees Kelvin, which can never be actually reached. Another implication of the Carnot engine is that it requires that heat be drained out of the cylinder for it to work. If the heat were not drained, pushing the piston back in would require as much work as was produced when it was pushed out, and the net work delivered by the engine would be zero.

- The Carnot engine is a theoretical construct, devised simply to provide a starting point for thermodynamic analysis and the understanding of practical engines. The details of operation of practical engines don't quite match this ideal due to factors such as friction, heat losses, the temperature limits of the materials used to build the engine, and so on. However, the basic revelations of the Carnot engine hold true: efficiency is roughly proportional to the temperature difference between engine inlet and outlet, and waste heat must be produced by the engine for it to work.

The fact that engine efficiency is directly proportional to the temperature difference gives an insight into the definition of entropy as heat transfer at a given absolute temperature. As the absolute temperature falls for a heat transfer process, the ability of the heat to do useful work, its "quality", falls or "degrades" as well, or in other words its entropy increases. A quantity of heat transferred at a high temperature has low entropy; the same quantity of heat transferred at a low temperature has high entropy.

This is the Second Law of Thermodynamics in action. Over time, in a closed system the transfers of heat occur at lower and lower temperatures, reducing the ability of the system to do work for both artificial engines and natural processes. This implies that the entire Universe is running down slowly, towards an ultimate "heat death" in the distant future.

PRACTICAL ENGINES

- A number of practical heat engines have been invented and are now in use. One of the most elegant, though one of the least-used, is the "Stirling cycle" engine. This is an "external combustion" engine, meaning that it is powered by an external source of heat. Any reasonably practical heat source can be used, such combustion of gasoline, kerosene, wood or manure, or sunlight focused by a mirror. All the Stirling cycle engine needs to run is a source of heat on one side and a cooling sink on the other.

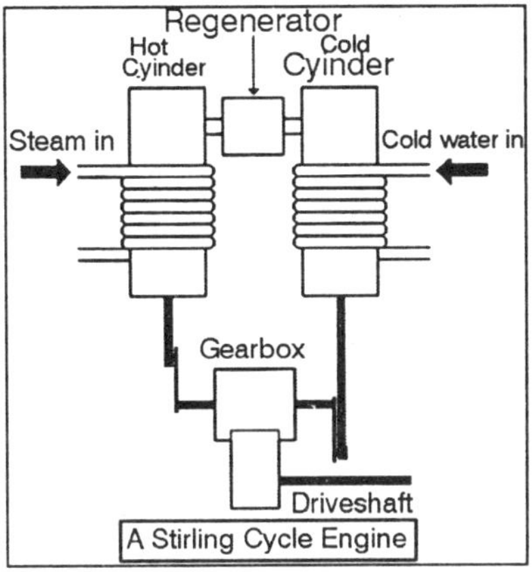

A Stirling Cycle Engine

The Stirling cycle engine's indifference to fuel source is its main advantage, and it is of some use in undeveloped countries where access to fuels is limited. Its disadvantage is that it has a poor "power to weight ratio (PWR)". All other engines in common use can provide the same power for much less weight.

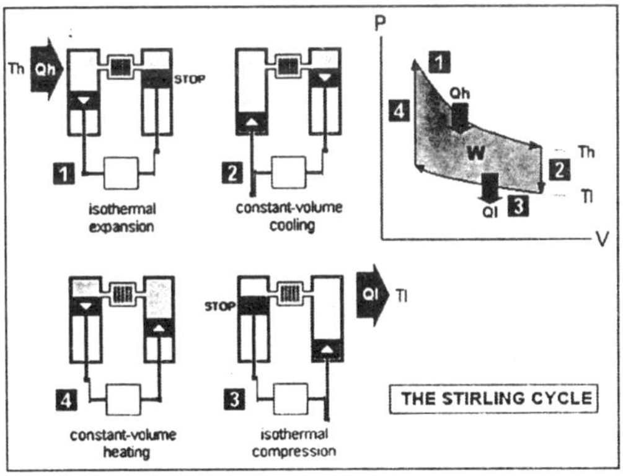

THE STIRLING CYCLE

The illustration below is an example of a simple (and idealized) implementation of a Stirling cycle engine. It consists of two cylinders, each with their own piston, linked through a gearbox. In this example, steam is fed around the first "hot" cylinder as a heat source, while cooling water is fed around the other "cold" cylinder as a heat sink. The "heads" of the two cylinders are linked by a heat-storage element known as a "regenerator".

The operation of the engine's gearbox is a bit difficult to follow, since it will bring one piston to a dead stop while moving the other piston, or allowing it to move.

The gearbox steps the engine through a four-part cycle as follows:

- While the piston in the cold cylinder is stopped at the top of its travel, heat input into the hot cylinder causes the piston there to move downward. This is the power stroke, an isothermal expansion process.
- The piston in the hot cylinder is then moved to the top of its travel while the piston in the cold cylinder is moved to the bottom of its travel. The volume of the working gas in the engine remains the same, while heat from the gas is stored in the regenerator. This is a constant-volume cooling process.
- The piston in the hot cylinder is stopped at the top of its travel while the piston in the cold cylinder is moved up the cylinder to half its full travel. This compresses the working gas, but the temperature remains constant because heat is being drawn off by the cold water. This is an isothermal compression process.
- Now the piston in the hot cylinder is moved downward to half its full travel, while the piston in the cold cylinder is moved to the top of its travel. This is a constant-volume process, in which heat from the regenerator is fed back into the gas flowing into the hot cylinder. The cycle is now ready to begin all over again.

This two-cylinder configuration is sometimes known as an "alpha" Stirling engine. In practice, Stirling engines are often built in a "beta" or "displacer" configuration, in which both pistons are in one cylinder, on top of one another and operated by concentric rods, and the regenerator runs up the sides of the cylinder, with ports between the two pistons and above the top piston. The top piston is called the "displacer", since its function is to "displace" the working fluid, while the bottom piston is called the "power piston", since it provides the drive power out of the engine.

The beta Stirling engine's operating principles are exactly the same as those of an alpha Stirling engine, though it is trickier to understand. There is also a "gamma" Stirling engine configuration, which is exactly the same as the beta configuration except that it is rearranged a bit to avoid the use of concentric rods. The illustration below shows a beta Stirling engine stepping through its cycles, with an idealized Stirling engine shown below to help clarify its operation, along with a gamma configuration engine.

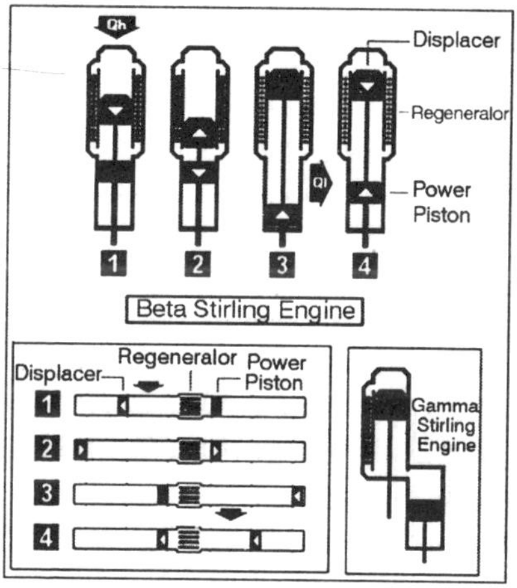

In practice, the gearing system that drives a beta Stirling engine is designed so that a piston may slow down but won't ever actually stop, except for the instant between reversals of direction. This construction "rounds off" the edges of the PV diagram, but otherwise the operation is the same.

- The most popular heat engine in service is the automotive internal combustion engine, known formally as the "Otto cycle" engine and more popularly as the "four-stroke" engine.

Operation of the Otto cycle is conceptually simple:

- In the first part of the cycle, the "intake stroke", the intake valve opens and the piston moves down to its lowest position. This draws in a gasoline-air mixture at (constant) atmospheric pressure.

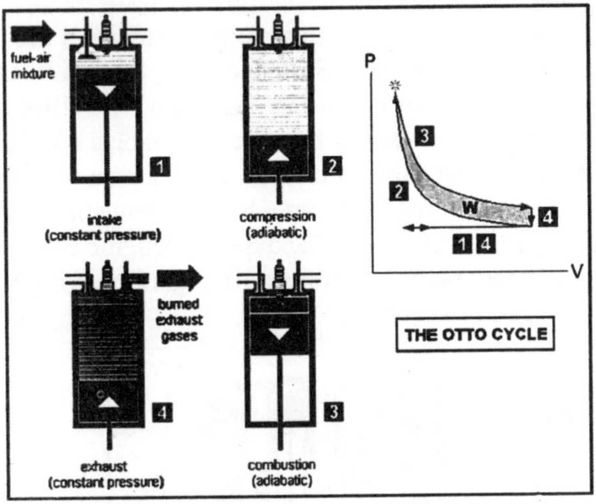

- In the "compression stroke", the intake valve is closed and the piston moves to its top position, compressing the fuel-air mixture. This is done quickly enough so that this part of the cycle is effectively adiabatic.
- In the "power stroke", an electric spark plug ignites the compressed fuel-air mixture, driving the piston to its lowest position.
- In the "exhaust stroke", the exhaust valve opens and the piston moves up, exhausting the combustion products of the fuel-air mixture at (constant) atmospheric pressure.

Analysis of the Otto cycle engine shows that its efficiency is mostly dependent on the "compression ratio", that is, the ratio of maximum compression of the fuel-air mixture to atmospheric pressure. The greater the compression ratio, the more efficient the engine. However, the Otto cycle engine is limited on the level of compression it can obtain, since at high compressions the temperatures and pressures will cause the fuel-air mixture to ignite spontaneously before the piston reaches the top of its travel. This phenomenon is known as "engine knock".

The "Diesel cycle" engine avoids this problem. It has the same general four-stroke operational cycle as the Otto cycle engine, but uses a fuel injection system to spurt fuel directly into the cylinder at the end of the compression stroke, permitting higher compression ratios. The Diesel engine uses heavier fuels than the Otto cycle engine. It does not use spark plugs, instead using "glow plugs" that are heated and assist ignition of the fuel-air mixture when the compression reaches the proper level.

A second variation on the Otto cycle is the "two-stroke" engine, in which the intake and expansion cycles are combined, as are the compression and exhaust cycles. This means that the two-stroke engine has, in the limit, twice as much power for a given RPM than a four-stroke engine and is correspondingly lighter.

The problem with two-stroke engines is that they unsurprisingly burn very dirty, and so have been generally banned by air-pollution regulations. However, experimental two-stroke engines have been built that use electronic control systems to meet air-pollution regulations.

- There is a very wide range of configurations of four-stroke and two-stroke engines air or water cooled; or pistons in flat, inline, vee, or radial configurations. There are also exotic variations on the theme, such as the well-known "Wankel" rotary engine, which instead of a piston in a cylinder uses a rotor in the form of a "fat" triangle rotating inside a combustion chamber with a cross section like that of a kidney bean.

  Describing all the details and variations would be a major document in itself; it should be enough to say here that no matter what the configuration is, all such engines work on the same basic principles.

One configuration, the "free piston engine", is so unusual that it is worth adding here. The name is due to the fact that this class of engine doesn't have any driveshaft: the pistons just pop back and forth and down without being connected to anything, being driven forward by combustion to compress a closed-off air space and being driven back again. This sounds pointless, but the motion of the pistons can be used directly to pump gases or liquids.

The diagram below shows a free piston engine used as a water pump. Its basic operation uses the four-stroke Diesel cycle; there is no reason in principle that it couldn't use a four-stroke Otto cycle or a two-stroke cycle instead. If a free piston engine were used to pump gases, say to drive a jackhammer, it could be made simpler, with the air pumped into the combustion chamber and the gas driven out the engine exhaust. The advantage of a free piston engine is obvious: it is much simpler and uses fewer parts than a comparable Diesel engine driving a separate pump.

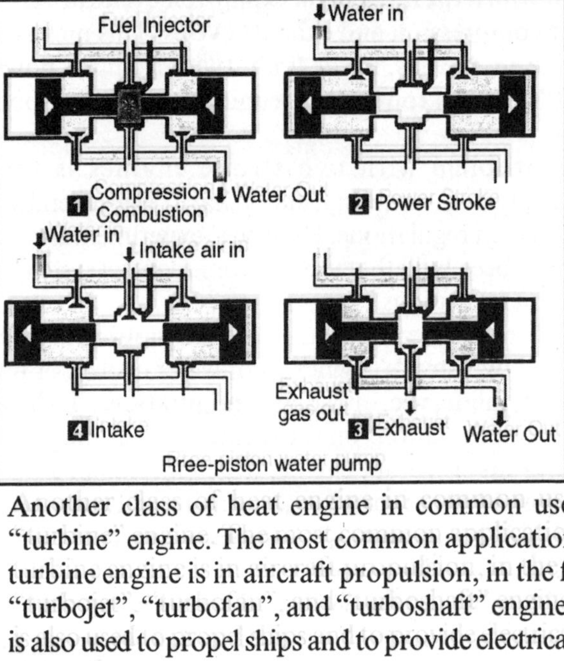

Rree-piston water pump

- Another class of heat engine in common use is the "turbine" engine. The most common application of the turbine engine is in aircraft propulsion, in the form of "turbojet", "turbofan", and "turboshaft" engines, but it is also used to propel ships and to provide electrical power generation.

Turbine engines are generally known as "Brayton cycle" engines. A turbojet is an "open" Brayton cycle engine, in that it operates on a continuous flow basis rather than in a loop, as follows:

- Intake air is first compressed by a "compressor" assembly. This is an adiabatic process.
- Fuel is then mixed with the compressed air and burned in the "combustion chambers" of the engine. This is a constant-pressure process.

- The hot gases generated in the combustion chamber flow out the back to provide reaction thrust, and also spin a "turbine" to run the compressor.

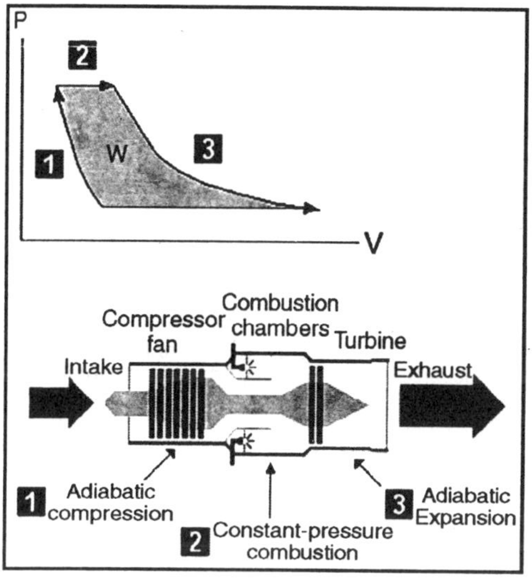

A turbojet is a little different from the other engines considered in this section because it is mainly intended to generate thrust. Turbine engines used to power ground vehicles, "turboprop" engines used to spin an aircraft propeller, and "turboshaft" engines used to spin a helicopter rotor mainly generate torque.

"Closed" cycle turbine engines are used in powerplants and on large vessels. They may operate much like aircraft turboshafts, or they may be driven by compressed steam heated by an external thermal source. The steam then drives a turbine to provide power, and is finally cooled to be routed around to the intake of the compressor again. Closed cycle engines include secondary "heat exchangers" to draw some of the heat out of the exhaust of the turbine and preheat the input, improving energy efficiency.

- One elegantly simple example of an open-cycle turbine-based heat engine is a solar power plant that may be built in the hot deserts of Australia. The design looks something like a big chimney, as tall as a skyscraper, that contains a set of turbines. The chimney is open at the bottom and is surrounded by a circular greenhouse-type structure in which the ceiling rises from the rim to the centre. The sun heats up the air in the greenhouse; the hot air flows inward and then into the openings in bottom of the chimney, and rises up the chimney into the cooler air at the top, driving the turbines.

POWER TURBINE SYSTEM (GVG / PD)

What is particularly interesting about this scheme from the point of view of elementary physics is that it illustrates how a temperature difference, in this case between the bottom and top of the chimney, is required to allow the engine to work.

If it wasn't, the air at the bottom of the chimney would not rise and the turbines would stay idle. In addition, the greater the temperature difference the hotter the air at the bottom of the chimney compared to the air at the top the faster the airflow up the chimney, allowing more power to be obtained out of the turbines.

- Another interesting and unusual engine from a physics point of view is the "dipping bird" toy. This item consists of a bird with a glass body made of a tube with a head and beak on one end and a relatively large bulb on the other. The body pivots on a two-legged stand. The bulb is filled with a fluid that vaporizes easily at room temperature.

To get the dipping bird to work, a glass of water is set in front of it and its beak is placed in the water. The bird will pop upright and stay there for a while; then tip over, put its beak in the water, and remain there for a while; and will then pop upright again, repeating the cycle until the water supply goes dry. What happens is that, while the bird is upright, the volatile fluid in the bulb in the tail evaporates and rises to the head.

This upsets the bird's centre of gravity and it pivots over, dropping its beak into the water. The water cools the fluid and it flows back down to the bulb incidentally, the bird's body never goes horizontal or below horizontal. This shifts the centre of gravity back and the bird goes upright again.

The dipping bird obtains its energy from ambient heat, but it needs a cold sink, the glass of water, in order to work. Of course, the bird won't dip if the ambient temperature is too cool to allow the fluid to vaporize, and it won't right itself if the ambient temperature is too warm to allow the fluid to liquify.

This is not a practical engine, though in the 1970s a humour magazine, reflecting the early energy conservation fad of the decade, envisioned giant dipping birds operating as central power stations, and even drew up blueprints. However, the same principle has been considered for practical applications.

In the early 1990s, work was done on the development of torpedo-sized roving robot submarines for oceanographic studies. These machines known as "Slocums" for Joshua Slocum, who in 1898 was the first person to sail around the world solo are powered solely by temperature differences in the oceans.

A Canadian engineer named Doug Webb came up with the idea for the powerplant. He had been working on a sea float that could control its floatation. The float contained an oil reservoir and an external bladder: the oil was pumped out of the float into the bladder to make the thing rise, and pumped back in to make it sink. This was a simple enough idea, but Webb kept thinking it over and coming up with refinements.

First, he decided that he might not really need to use a pump, he could just shift oil around by heating and cooling it. Next, he realized that the temperature differences at different depths in the ocean could do the heating and cooling for him. Finally, he realized that if he put "wings" on the float, it would be able to control its direction as it sank or rose.

The thermal engine that Webb finally came up with hardly

looked like an engine at all. The engine components were contained inside a cylindrical pressure casing that made up a body section of the Slocum, except for an inflatable bladder mounted outside the cylinder in the Slocum's frame.

The bladder could be filled with liquid glycol (normally used as automotive antifreeze) and was connected to a matching one inside the cylinder through a tube. The cylinder also contained a pair of pressure tanks, one containing nitrogen and a bellows filled with liquid glycol, and the other containing a special pure hydrocarbon. The components were linked by a system of valves, as shown below:

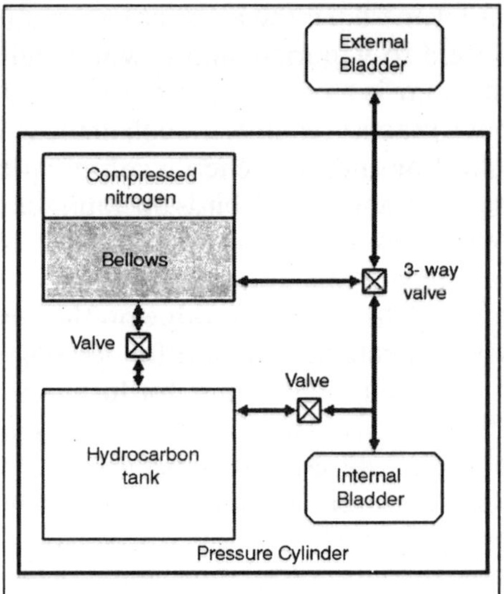

Webb's Thermal Engine

The scheme for operation of this device was devious. The hydrocarbon was liquid at about 15 degrees Celsius and solid at about 5 degrees Celsius, with the transition between the two at about 10 degrees Celsius. Solidified, the hydrocarbon resembled candle wax. On the surface, the external bladder would be filled with glycol; the internal bladder was empty; the hydrocarbon was liquid and the hydrocarbon tank was full; and the bellows was full of glycol. In this state, the Slocum floated. To make it sink, the 3-way valve was opened, causing the glycol to flow into the internal bladder.

At the bottom of the Slocum's descent, a kilometre and a half under the sea, the submarine was cold enough to cause the hydrocarbon to freeze. This caused it to shrink by 10 per cent, with the evacuated space drawing off the glycol from the internal bladder through a valve.

Wind Turbine (GVG/PD)

To ascend, the 3-way valve was closed to connect the bellows and the external bladder, and the compressed nitrogen then drove the glycol into the external bladder, causing the Slocum to rise. This done, the 3-way valve was closed again, and the valve between the bellows and the hydrocarbon tank opened. As the Slocum rose, the hydrocarbon expanded, forcing the glycol back into the bellows and compressing the nitrogen. By the time, the Slocum had reached the surface and was ready for another dive.

The engine could operate effectively under the pressures of the deep sea. No fuel was required, though the Slocum did need battery power to activate the valves. It wasn't a very powerful engine: in practice, it could propel a 40 kilogram Slocum no faster than a kilometre per hour.

That was still fast enough to allow the Slocum to travel across the seas over a period of months or years. Since the Slocum was supposed to operate unattended for five years in a harsh oceanic environment, the diving fins were fixed at an angle for simplicity, with the Slocum adjusting its orientation by shifting the battery pack inside the pressure cylinder.

- Webb got his thermal engine to work, and development of Slocums has continued. A prototype named "Spray" was sent on a journey from Greenland to Spain in the spring of 2006, and plans were made for mass production of the vehicles. It is certainly fun to see such a clever way of getting something for seemingly nothing.

Many forms of renewable energy systems wind power and hydropower have a similar sort of appeal. Obviously, none of these machines actually violate the law of conservation of energy, they just make use of energy sources available in the ambient environment, with the energy inputs provided ultimately by the light and heat of the Sun. Trying to trace out the engine cycles for wind power (global weather) and hydropower (global rain cycle) would be an interesting exercise.

REFRIGERATION SYSTEMS

- A heat engine taps the flow of heat from a hot source to a cool sink to do work. It is also possible to reverse this flow, using work to remove heat from a cool source. This is the fundamental principle of how a refrigerator (or air conditioner) works.

In practice, a simple "refrigeration system" consists of a long sealed pipe arranged in a loop. Half of this loop is placed inside the refrigerator while the other half is placed in the environment outside the refrigerator, where the heat from the inside will be dumped. Both halves of the loop of pipe are turned back and forth with a series of hairpin turns to make them more compact, forming what are referred to as "coils", though they only coil in two dimensions, not three as in a coil spring. The half of loop inside the refrigerator is called the "evaporator coil" or just "evaporator", while the part outside the refrigerator is called the "condenser coil" or just "condenser". There are of course two connections between the evaporator and condenser. One connection is linked by an electric compressor, while the other is linked by a "flow restrictor" or "expansion valve".

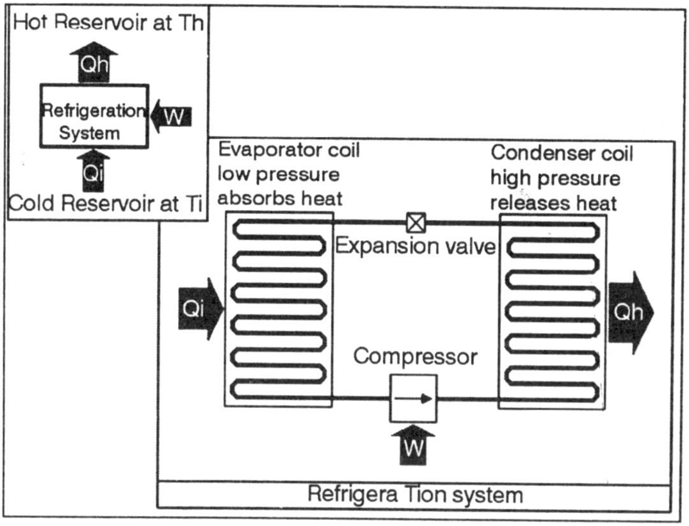

The entire loop is filled with a "working fluid" or "cooling fluid" that changes easily from liquid to gas. The compressor drives the cooling fluid from the evaporator into the condenser, while the expansion valve throttles the flow of the cooling fluid from the condenser back into the evaporator.

The cooling fluid evaporates as it goes from the condenser through the expansion valve into the evaporator, and becomes cooler due both to its change of state and its expansion in volume.

The cool gas in the evaporator soaks up heat from inside the refrigerator. The gas is then condensed and liquified when the compressor drives it from the evaporator back into the condenser. The warm fluid releases heat to the environment, and circulates back to the expansion valve to begin the cycle again. For more aggressive cooling, a "two-stage" refrigeration system can be used, consisting of two such refrigeration loops in cascade.

- In the earliest cooling systems, concentrated ammonia was used as the cooling fluid. It worked well from a thermodynamic point of view, but it had a major drawback: it was toxic. There are stories, apparently true, that people were actually killed by refrigerator leaks.
  This led to the introduction of a new class of cooling fluids, a set of compounds based on chlorine and fluorine

known as "chlorinated fluorocarbons (CFCs)". They looked like an absolutely perfect solution to the problem, since they not only had nearly ideal thermodynamic properties, but they were almost completely inert, non-toxic, and relatively cheap.

In the 1980s, however, atmospheric researchers began to discover that trace amounts of CFCs were accumulating in the upper atmosphere, where they were broken down by solar radiation. The chlorine in the molecules was able to cause the breakdown of "ozone", the O3 molecule, which covers the Earth in a high-altitude layer that is opaque to harmful radiation.

While ozone breakdown was only observed in the cold air above Antarctica, the possibility that increasing accumulations of CFCs could entirely wipe out the Earth's ozone layer destroying the shield that protects the planet from high-energy radiation was frightening enough to lead to international agreements to phase out the use of CFCs in refrigeration and air conditioning systems, as well as in their common use in plastic foam insulation.

There is a popular misconception that CFCs were banned because they were toxic. Actually, the truth was that they were too inert for their own good. Once they were released into the environment, they didn't degrade until they floated to high altitudes where they could do the most harm. CFCs are now being replaced by "hydrofluorocarbons (HFCs)" that don't contain chlorine, and by "hydrochlorofluorocarbons (HCFCs)" that do contain chlorine but break down before they reach high altitudes. Neither of these compounds are quite as efficient as cooling fluids or insulators as CFCs, however, so the conversion to HFCs and HCFCs had not been straightforward.

- An air conditioner works the same way as a refrigerator, pulling heat out of a room with the evaporator and dumping it into the outside environment with the condenser. If the flow of cooling fluid through the refrigerator was reversed, the system would then pull heat out of the environment and dump it into the room, heating the room. This is the basic principle of the "heat pump", a device in which the flow of cooling fluid can be reversed to either cool or heat a room.

Heat pumps are efficient compared to most other forms of heating, but only in fairly moderate climates where the temperature changes are not too drastic.

- An even more interesting reversibility is that if a Stirling cycle engine is run "in reverse", with the device being driven instead of producing work, it makes a very nice cooling system. In fact, while the Stirling cycle engine remains something of an oddity long after its invention, the "Stirling cryocooler" is in increasing if somewhat specialized use as a compact refrigeration system.

The basic principle of the Stirling cryocooler is exactly the same as that of a more conventional refrigeration system: compress a working fluid in the external environment and allow it to cool to ambient temperature, then transfer the working fluid inside the refrigerator enclosure and allow it to expand as a gas.

This is done with exactly the same four steps as the Stirling engine cycle, just implemented differently, with the "hot" cylinder becoming the "cold" cylinder, and the reverse:

- In the expansion phase, the piston in the cold cylinder, which is inside the refrigerator enclosure, is drawn out to cool the gas, rather than being driven by the expansion of the gas due to the input of heat. The cool gas then absorbs heat.
- The cool gas is transferred to the hot cylinder in the external environment in a constant-volume process, with the gas cooling the regenerator.
- The piston in the hot cylinder compresses the gas, with the heat due to compression leaking off into the ambient environment.
- The cooled compressed gas is shunted back to the cold cylinder in a constant-volume process, being given additional cooling by the regenerator in the process.

As noted in a previous chapter, a warm object will emit infrared radiation. An infrared camera can obtain an image of an object just from its infrared radiation, even if there is no other light source available. The military is particularly fond of such infrared camera systems, which they call "forward looking infrared (FLIR)" imagers. The military also uses "heat-seeking" missiles, such as the well-known "Sidewinder" missile, that will "home in" on a target aircraft from its thermal infrared radiation using a heat-seeking sensor. However, also as noted, both FLIRs and heat-seeking sensors suffer from an inherent constraint: they can't sense a target that is cooler than they are, and so it is desireable to cool the FLIR or heat-seeking sensor to improve its sensitivity. Early military FLIRs used in the 1960s had bulky refrigeration systems and had to be carried in fairly large aircraft. Early versions of the Sidewinder missile didn't have any cooling system, and so they could only be targeted on an aircraft's hot exhaust.

In the 1970s, improved versions of the Sidewinder were introduced that could be fitted before a mission with a can of compressed nitrogen gas, with the gas leaking out a pinhole to provide cooling. These improved "all aspect" Sidewinders could lock onto an aircraft target from any angle and no longer had to be targeted on an aircraft's exhaust. The limitation of this scheme was that the cooling system only worked for a relatively short time.

The introduction of compact, relatively low-cost Stirling cryocoolers in the 1990s provided a better solution, since such a device can provide cooling for the sensor for as long as power is available.

FOOTNOTE: MAXWELL'S DEMON AND THE SECOND LAW

- As noted previously in this document, the law of conservation of energy rules out most perpetual motion machines, the "perpetual motion machines of the first kind", but there are also "perpetual motion machines of the second kind" that *don't* violate energy conservation. Such machines operate in defiance of the Second Law, for example by attempting to extract energy from ambient heat without even the implied presence of a temperature difference.

A typical example of such a machine may incorporate a screen that, so the inventor claims, will allow fast-moving molecules to pass through and block slow-moving molecules, resulting in the buildup of a temperature difference between the two sides of the screen that could then do useful work. The screen might have tapered passages, with large holes on one side of the screen narrowing to small holes on the other, which supposedly will tend to pass more fast-moving molecules than slow-moving ones.

Such schemes still end up being busted by the physics police: the "magic screen" described above is a fantasy, just as absurd as Maurits Escher's circular waterways. In practice, even if there were originally a temperature difference across the screen, it would quickly balance out. As SCIENTIFIC AMERICAN magazine put it, in a column on rejection letters written as Japanese haiku poems:

- Though the specs look nice,
- A stop sign halts your machine.
- Thermo's second law.

However, there are also a number of physicists who like to play the perpetual-motion game as a thought experiment to see what interesting insights it may provide. Nobody sensibly thinks for an instant that there is any way to get around the law of conservation of energy, but could some machine be invented that exploits a loophole in the second law of thermodynamics? The most famous of these theoretical perpetual motion machines is "Maxwell's Demon", originally devised by James Clerk Maxwell, which amounts to a "smarter" approach to a magic screen. In this scheme, the system consists of two chambers containing gas molecules. The chambers are separated by a door that is opened or shut by a tiny "demon". When the demon spots a fast-moving molecule in the first chamber, it opens the door to let it into the second, and keeps the door closed for slow-moving particles. In this way, the demon can build up a stockpile of energetic molecules that can be released or otherwise used to do useful work.

While obviously there is no demon available to do the job, it was still hard to understand why such a scheme could not be used in principle to build perpetual-motion machines, and through the invention of the demon Maxwell was posing the question to later generations as to why it wouldn't work. Incidentally, Maxwell did not name the demon after himself; it was named by others later in honour of Maxwell. In 1929 by the Hungarian physicist Leo Szilard proposed a solution. Szilard asked: how does the demon know which molecule is moving faster than the others? Basically, he'd have to "turn on the lights" in the chamber so he could see the molecules, and even if he only used a selective beam of light to do so, Szilard showed that the energy expended in sensing the motion of the particles would always be greater than the amount of energy gained by sorting the molecules out. The Second Law wasn't violated.

However, Szilard came up with a simple variation on Maxwell's Demon, sometimes called "Szilard's Demon", that evaded this obstacle, taking things more or less back to square one. Eventually studies of information theory showed that there was a certain unavoidable amount of energy expenditure in making a decision, and this amount of energy was still enough to guarantee the Second Law.

- There have been a number of variations on the theme of Maxwell's Demon. An interesting one was devised by Dick Feynman, who enjoyed being clever and unorthodox. His demonic "motor" was envisioned as a microscope gear connected along its axis to a small paddle wheel. The gear was ratcheted so that it can only move in one direction. If the Feynman motor was placed in a gas or fluid, molecules would strike the paddle wheel. Since the gear could only move in one direction, the paddle wheel could only turn one way, and useful work could be obtained from random (Brownian) molecular motion. Feynman figured that his motor could lift fleas.

Having presented this idea, Feynman then demolished it. The weak point was the spring-loaded ratchet. Thermal action would also cause the ratchet to bounce out of place, and the skewed configuration of the gear teeth would make it more likely the gear would move in reverse if that happened. The jerk back through the paddle wheel as the ratchet snapped back into place would dump heat back into the medium.

Feynman's motor was a theoretical toy, but others have taken it a bit more seriously. In 1997, researchers at Boston College actually built something like a Feynman motor, using benzene molecules (six carbon atoms in a hexagonal ring) to build a three-blade paddle wheel with a ratchet.

Chapter 10

Waves and Sound

One of the properties of matter is that is can support transfers of mechanical energy, whether the matter is solid, liquid, or gas, without any net movement of the molecules involved. Such transfers are referred to as "wave motion", or a transitory displacement of atoms within the matter that passes on kinetic energy. This chapter discusses the physics of mechanical wave motion. Light also involves wave motion, but not of a mechanical sort, and will be discussed in the following chapter.

WAVE MOTION / STANDING WAVES

- We are all familiar with wave motion. The sounds that we hear are waves passing through the atmosphere, with the air compressing and decompressing in an oscillating fashion, passing the sound energy outward at hundreds of kilometers per second. We are all familiar with the waves that expand outward after we drop a big rock in a pond, or the waves that we create by whipping a rope tied to a fence up and down.

There are two kinds of waves: "transverse", and "compression" or "longitudinal" waves. In a transverse wave, the molecules move back and forth at a right angle to the direction of motion. A rope being shaken with an up-and-down motion generates a transverse wave, and the ripples we see spreading across a pond are also transverse waves. In a compression wave, the molecules move back and forth in the direction of the wave motion. Sound is a compression wave. A more easily visualized compression wave can be produced by taking a "Slinky" toy, holding it in both hands, and shaking it with a side-to-side motion.

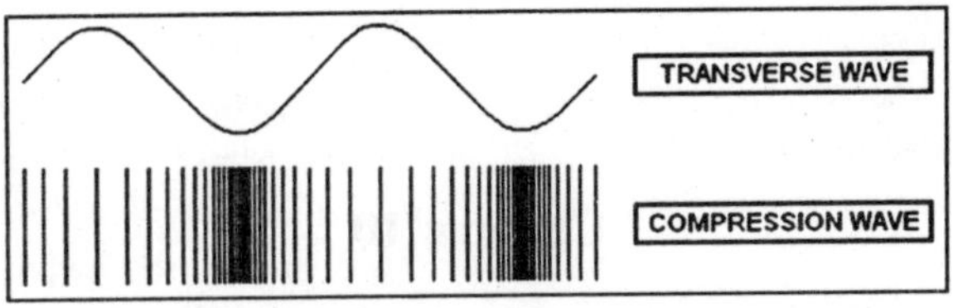

Transverse waves can only arise in a solid medium, or at the interface between a liquid and gaseous medium. A strictly gaseous or liquid medium can only support longitudinal waves.

- Waves have a specific speed of propagation that depends on the matter transmitting the wave, or "wave medium", and the environmental conditions. In dry air at sea level at freezing temperature (zero degrees Celsius), the speed of sound is 331.6 meters per second, or 1,194 KPH.

In reality, the speed of sound is far from constant. For one thing, an increase in temperature increases the speed of sound. For example, at 20 degrees Celsius, the speed of sound increases to 344 meters per second or 1,238 KPH. At higher temperatures, the molecules are moving more quickly and can transmit sound energy more quickly.

Moisture in the air also increases the speed of sound, since the water molecules are on the average lighter than the air molecules they replace, and lighter molecules have less inertia and react more quickly. In addition, the speed of sound also slows with altitude, since the air is less dense and the molecules cannot transmit energy to each other as quickly. The speed varies inversely to the square root of the density.

In general terms, then, the speed of sound:

- Increases with temperature.
- Decreases as the size of the molecules involved grows greater.
- Increases with density.

Sounds generally move much faster in dense liquids and solids than in gases. However, in solids sounds are also dependent on the elasticity of the material, increasing with elasticity. The speed of sound in water is about 1,525 meters per second or 5,490 KPH at normal temperatures, increasing greatly with temperature. Of course, sound is a compression wave, and this is not the same as the velocity of the transverse surface waves that travel across bodies of water, which surfers can "ride" for their amusement.

Surfing would not be fun for most people if the speed of such waves was any real percentage of 5,490 KPH. The speed of sound in copper is about 3,353 meters per second or 12,070 KPH at normal temperatures, but oddly decreases with temperature, because the metal's elasticity falls off.

- Waves can have different "amplitudes", or variation in magnitude between "peaks" and "troughs" of the wave, with the power carried by the wave increasing with amplitude. The energy carried in a wave is proportional to the square of the amplitude. It should be noted that in many of the examples of mechanics presented earlier in this document, energy losses were described as due to friction, but that's a bit of an oversimplification; energy is also carried off by sound waves.

Waves can have different spacings, or "wavelengths", between successive regions of maximum density (for a compression wave) or peaks (of a transverse wave). The wavelength is related to the "frequency" of the wave, or the number of times it oscillates in a second while passing through a fixed location, by the simple relationship:

frequency = velocity / wavelength

For example, for a sound wave propagating at 344 meters per second and a wavelength of ten centimeters, the frequency is 3,440 "cycles per second" or, using modern terminology, "hertz (Hz)". For a wavelength of 20 centimeters, the frequency falls to 1,720 Hz. Waves correspondingly have a "period", which is the time between passages of consecutive crests troughs of the wave. The period is given by the inverse of the frequency:

period = 1 / frequency = wavelength / velocity

Finally, waves have a "phase", in that the peaks of one wave may not coincide in time with the peaks of another wave of the same frequency.

Suppose Dexter has two wave tanks, side by side, and generates waves of the same frequency in both. At first the two wave generators are set up to operate in perfect synchronization, but then one is momentarily paused or delayed so that the peaks of its waves don't match up with the other. The two waves are now "out of phase".Phase is measured in degrees or radians. When the two waves are in phase, they have a "phase difference" of 0 degrees. As one of the generators is delayed so that the phase falls behind that of the other, the phase difference increases.

When the two waves are completely out of step or "antiphase", with the peaks of one coinciding with the troughs of the other, the phase difference is 180 degrees. If the generators is delayed again, its wave output will eventually come back into phase with the other, and the "phase shift" will have gone through a full cycle of 360 degrees. Please remember that phase is a comparative concept. One wave by itself doesn't really have a phase, but multiple waves being compared, or a individual waves compared to a specific reference time, do have phase.

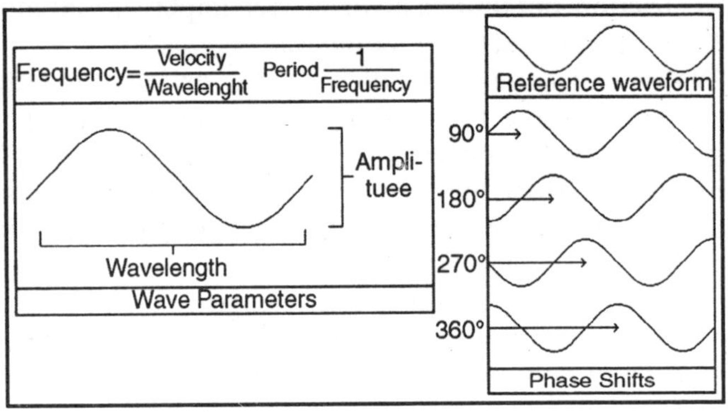

- The most recognizable, and for purposes of mathematical analysis fundamental, form of a wave is a "sine wave", which is produced by the trigonometric sine function. It is the wave which is produced if the position of the tip of a clock hand moving at a steady rate is plotted over time. Given a sine wave of a given amplitude, velocity, and period, then that wave can be represented by the formula:

Amplitude × SIN(360 × time / period) or equivalently: amplitude X SIN(360 × time × frequency) amplitude × SIN(360 × time × velocity / wavelength)

If a phase needs to be added, it can be expressed as: amplitude × SIN(360 × time / period + phase)

ACOUSTICS AND MUSICAL SOUND / SONAR

- Sound waves are of course important to humans for speech communications and music, and the study of sounds is a domain in its own right, known as "acoustics".

The sounds we hear can be characterized by three parameters: the loudness of the sound, its "pitch", and its "quality". The loudness of the sound obviously depends on the level of energy being pumped into the sound waves, but interestingly our perception of loudness is roughly proportional to the cube root of the energy of that sound. In simpler terms, to make a sound seem twice as loud, we have to pump eight times the energy into it.

The loudness of sounds is measured in a logarithmic scale known as "decibels", defined as ten times the exponent of 10 for the loudness value. The threshold of hearing is defined as 0 decibels. 10 decibels is ten times as loud, while 20 decibels is 100 times as loud, and 30 decibels is 1,000 times as loud. Typical sounds of traffic are about 70 decibels, while the noise of a nearby jetliner is about 140 decibels.

Pitch is simply the frequency of the sound. Somebody who talks in a deep gravel voice has a low-pitched voice, while most children tend to have high-pitched voices. The concept of pitch is of course fundamental to music, with musical "notes" of given frequencies arranged on a musical "scale". The classical scheme defining a musical scale involves dividing the range of musical tones into "octaves", named because each octave contains the eight whole notes "A", "B", "C", "D", "E", "F", "G", and "A", centered on the "Middle C" note at 264 hertz. Each "octave" of eight notes below Middle C halves the frequency, while each octave above doubles it, as follows:

– C'''	33 hertz
– C''	66 hertz
– C'	132 hertz
MIDDLE C:	264 hertz
+ C'	528 hertz
+ C''	1,056 hertz
+ C'''	2,112 hertz
+ C''''	4,224 hertz

There are different ways to define the tones in each octave. The most common, for Western music at least, is the "equitonic" scale. Examination of a piano keyboard shows that there are twelve keys in each octave, with seven white keys defining the "whole" notes and five black keys defining "sharp notes" and "flat notes". In the equitonic scale, the ratio from one of the keys to the next is the same for each key.

Since an octave means a doubling of frequencies, that means that this interval is the twelfth root of 2, or about 1.05946. The frequency values are rounded off to integers in practice. The following table gives the tones for the first octave above Middle C, with a "#" indicating a sharp note and a "-" indicating a flat note:

C	262	hertz
C#/D-	277	hertz
D	294	hertz
D#/E-	311	hertz
E	330	hertz
F	349	hertz
F#/G-	370	hertz
G	392	hertz
G#/A-	415	hertz
A	440	hertz
A#/B-	466	hertz
B	494	hertz
C'	524	hertz

Other scales, such as the more traditional "diatonic" scale, do not use equal intervals, but this makes changing to different keys more troublesome.

The quality of a sound describes the *kind* of sound we hear, regardless of tone. Most people can distinguish a note played from a piano from a note played from a flute, even if they're the same tone. This intuitively obvious, but not so easy to define in detail.

One of the important aspects of musical quality is the fact that, in general, a musical tone will not consist of a single frequency. This will be true for a simple electronic tone, but if Dexter presses the Middle C key on a piano, it will not only produce a sound with a frequency of 262 hertz, it will also simultaneously produce a range of other sounds of lesser amplitudes and higher frequencies, mostly integer multiples such 524 hertz and 1,048 hertz.

These different sounds are the "components" of the actual tone. The basic and most significant component of the tone, in this case 262 hertz, is called the "fundamental", while the component at the higher frequencies are called "harmonics". The set of these sounds is referred to as an "audio spectrum".

Different instruments generate different audio spectra for the same tone, which is one reason they sound distinct. There are other features to the quality of a musical tone, but further discussion would be leaving the domain of physics and crossing over into musical theory.

- The human ear can hear tones to up about 20 kilohertz. Most sounds in speech are below 3 kilohertz, so most telephones and voice radios only transmit sounds up to 3,000 kilohertz. Although, as the table above shows, musical tones don't go above about 5 kilohertz in themselves, loss of high frequency harmonics tends to make music sound muddy, and so "high-fidelity" systems need to produce sounds including frequencies of up to 16 kilohertz or better.

Some audio enthusiasts claim they can hear distinctions in sounds with tones above 20 kilohertz, but such claims rarely are validated by "double blind" tests, where neither the subject nor the person administering the test knows what the frequency makeup of the sounds are supposed to be. A few of the most earnest enthusiasts have responded to such negative results by attacking the validity of double-blind tests, but so far their arguments have not carried the day.

Some animals can hear well above the human range of hearing. For example, bats, whales, and dolphins have a hearing range several times that of humans. These animals use their vocalizations to perform ranging and location. They emit high-pitched sounds and then listen for echoes bounced off a target. The longer the interval between producing the sound and the return of its echo, the farther the target, and these animals also have ear structures that allow them to determine the direction of the echo.

Humans developed an electronic approach to the same procedure after the First World War to hunt submarines, ultimately naming the scheme "sonar", for "sound navigation and ranging".

REFLECTION, REFRACTION, AND DIFFRACTION / STANDING WAVES

- If a sound wave propagates through the air and runs perpendicularly into a flat, hard wall, it bounces off the wall and returns back in the direction it came from. This process is referred to as "reflection", and of course we recognize this as an echo.

This is the simplest scenario of a reflection, and in the real world it's a bit more complicated. First, if the flat wall is at an angle to the direction of the propagation of the sound, the sound will be reflected away at the same angle relative to the wall. For example, if the wall is at an angle of 45 degrees to the direction of propagation of the sound wave, the reflection will be at an angle of 45 degrees relative to the wall, or 90 degrees relative to the original direction of the sound. If the wall is made up of many different elements at many different angles relative to the propagation of the sound wave, it will generate reflections in many different directions, scattering the sound wave.

Not all the energy of the sound wave will be reflected from the wall; some of it will be absorbed by the wall as well. This combination of reflection and transmission will actually happen any time a sound wave makes a transition from one distinct acoustic medium to another, or in other words propagates through a "discontinuity".

- Another interesting phenomenon occurs when the wavefronts of a sound wave pass through a discontinuity at an angle. The wave will have different velocities in the different media separated by the discontinuity, and the portions of the wave that are propagating through the second medium do not travel at the same rate as those arriving from the first medium. This means that the direction of propagation of the wave in the second medium will be shifted from that of the first. This property is known as "refraction".

Yet another property of waves is their ability to flow around the edges of barriers. While a solid wall will block sound or water waves hitting it from one side, they will still slop around the edges of the walls, though their energy will be much diminished. This phenomenon is known as "diffraction". Neither refraction nor diffraction are all that important in the study of sound waves, but they are of great interest in the study of light waves, as discussed in the next chapter.

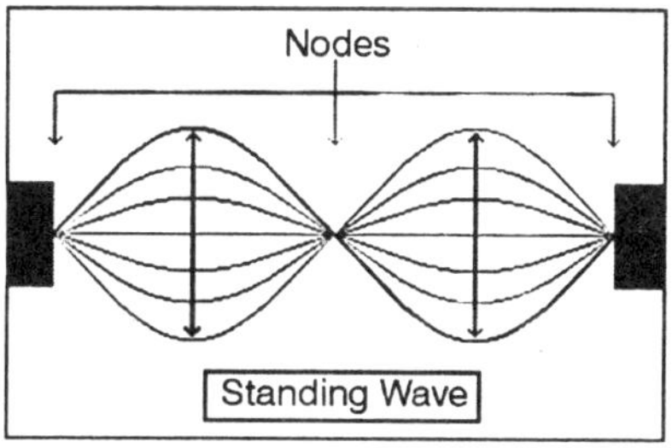

- If two waves of the same wavelength and amplitude travel through the same medium at the same velocity but in opposite directions, they set up "standing waves". The best-known example of this is that of a rope tied to a post. If Dexter shakes the rope up and down, the wave travels down the rope, hits the post, and then reflects back at him.

Dexter can time his shakes to synchronize in phase with the reflections so that fixed sections of the rope oscillate up and down, with the boundaries between the sections remaining effectively stationary. This is known as a "standing wave", as opposed to "traveling waves", or waves in normal unconstrained propagation. A standing wave actually consists of a superposition of two waves, one propagating in one direction and its reflection propagating back in the other direction. The stationary points on the wave are referred to as "nodes".

The standing waves on the rope will always be multiples of a half-wavelength to ensure synchronization between the input wave and the reflected wave.

The standing wave with a half-wavelength will have one node at each end and will move up and down in between; the standing wave with a full wavelength will have a node in the middle; the standing wave with 1.5 wavelengths will have two nodes in the middle; and so on. A plucked string on a musical instrument, such as a guitar, sets up standing waves.

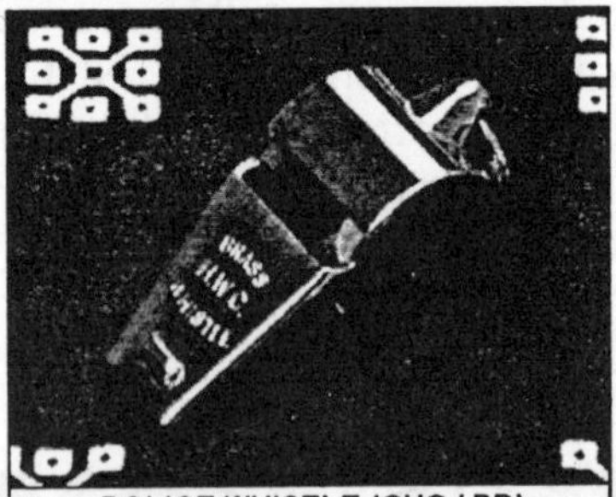

POLICE WHISTLE (GVG / PD)

Similarly, a police whistle generates standing waves with the wavelength given by the size of the whistle the smaller the whistle, the shorter the wavelength and the sharper the tone. The whistle acts as a "resonant chamber" or "resonant cavity" that selectively produces that tone from the air blown through it by the policeman.

There is generally a little ball in the chamber of the police whistle, which interferes with the resonance and breaks up the whistling tone, making it more noticeable; without the ball, a continuous whistling tone is generated.

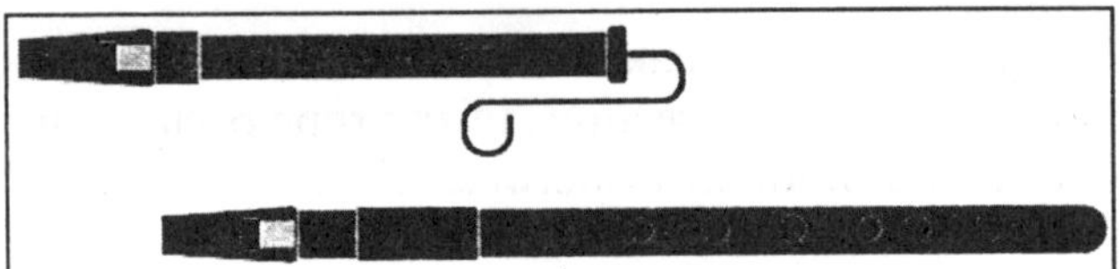

SLIDE WHISTLE & PENNYWHISTLE

A slide whistle allows the player to vary the tone continuously by moving a rod connected to a plug that varies the size of the resonant chamber. It is also possible to alter the resonant characteristics of a whistle by placing holes in the whistle tube, which the player can cover and uncover with his fingertips; such whistles include the classic Irish "pennywhistle" or "tin whistle", and the more traditional "recorder". Incidentally, pennywhistles and recorders are not technically referred to as "flutes". In a flute, the player blows across the top of a hole to generate the basic tones, instead of blowing straight through the instrument.

THE DOPPLER SHIFT

- The motion of a source emitting a wave affects the frequency of that wave. This phenomenon is known as the "Doppler effect". The most popular example of the Doppler shift is the sound of a moving train. If a train approaching at high speed blows a whistle, the frequency of the whistle is higher than it would be if the train were at rest. Similarly, if the train is moving away, the frequency of the whistle is lower.

This is easy to see mathematically. Suppose we designate the speed of sound as S and the velocity of an approaching train as V. Suppose the train is emitting a very short sharp blast on its whistle on an interval given by time T. If the approaching train emits a blast at time 0, then by the time it emits its second blast at time T, it will have moved closer by the distance V×T, and the sound of the second blast will not have to travel as far as the sound of the first.

The amount of decrease in the measured blast period will be given by the ratio of the train's velocity to the speed of sound. For example, suppose the blast period T is one second. If the train's velocity V is a tenth of the speed of sound, then in one second it will travel a tenth as far as sound does in one second, and the measured blast period Tm will be reduced to 0.9 seconds.

In more general terms:

$$Tm = T \times (1 - V/S)$$

In sum, the blast period observed from a moving train to the blast period observed from the stationary train is then:

$$Tm = T \times (1 - \text{fraction soundspeed})$$

If the train is moving away, the minus sign is changed to a plus, and the blast period is increased:

$$Tm = T \times (1 + \text{fraction soundspeed})$$

The blast frequency will be the inverse of the blast period.

This analysis applies to continuous waveforms as well. Suppose the train was also emitting a low-frequency audio signal in the form of a sine wave, and the blast was synchronized so that it was emitted at the beginning of each sine cycle. Clearly the two signals will remain in lockstep, and the sine wave will be Doppler-shifted exactly as the blast is.

WAVE INTERFERENCE

- One of the interesting features of waves is that their interactions are *additive*. For example, suppose Dexter generates two water waves of the same frequency and amplitude and have them meet head on. When two peaks pass through the same point, the amplitude of the peak will be doubled. When a peak and a trough match up, they cancel out or "interfere".

An interesting example of interference at work is to produce two sound waves of frequencies that are almost, but not quite, the same. As the two sounds move in and out of phase, they will eerily grow louder and quieter on a regular cycle. This is known as a "beat", though it has little to do with the "beats" that time musical compositions.

Suppose Dexter generates two sounds, one at a frequency of 300 hertz and the other at a frequency of 302 hertz. This means that they will both complete an integer number of cycles 150 and 151 respectively every half a second, and so the sound will vary in volume through a full cycle twice a second. The actual frequency of the sound heard will be 301 hertz.

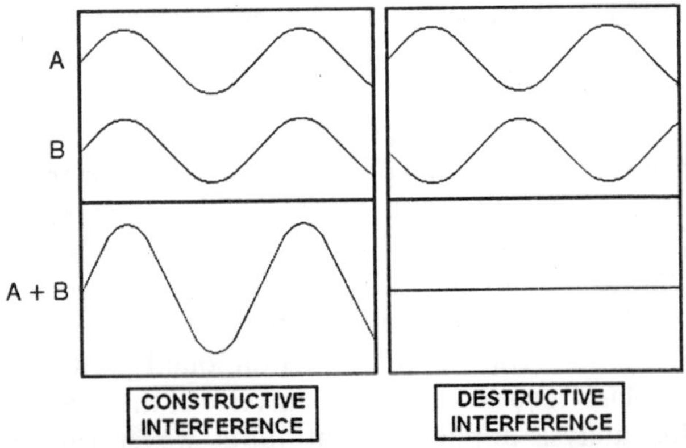

Wave interference is the basis of "active noise cancellation" systems. These systems include three elements: a microphone, a processor, and loudspeakers. The microphone samples external sounds and feeds them to the processor, which then calculates how to drive an antiphase signal to the loudspeakers that cancels out the external sound. Of course, this works best when the sounds are regular and predictable, for example the constant rumble of a large ventilation fan.

- One of the interesting implications of constructive interference is the concept of "bow shock". Suppose Dexter is at the helm of a boat sitting still on water. If he moves around in the boat, the rocking of the boat sets up ripples in the water that travel in a circle, radiating outward from the boat.

Now suppose Dexter starts driving the boat forward; it then sets up a continuous series waves that propagate circularly from every point that the boat moves through. If the boat is moving slower than the speed of the waves, this creates a Doppler shift, with the wavelength compressed in front of the boat and expanded behind the boat. The interesting thing is what happens when the boat exceeds the speed of the waves in water. All the waves in the front of the boat merge together, leaving behind a high-amplitude, vee-shaped bow shock wave behind the boat.

Water skiers are very familiar with this bow shock wave, since they like to hop over it or even surf on it for a while. The angle of the vee grows narrower as the speed of the boat increases. Supersonic aircraft also produce a bow shock wave that results in the propagation of an abrupt high-pressure wave in a conical footprint under the aircraft's flight path, causing a "sonic boom" that can rattle windows. Such sonic booms were common in the late 1950s and the 1960s, before the complaints of the citizenry forced the military to impose more stringent flight rules on their pilots.

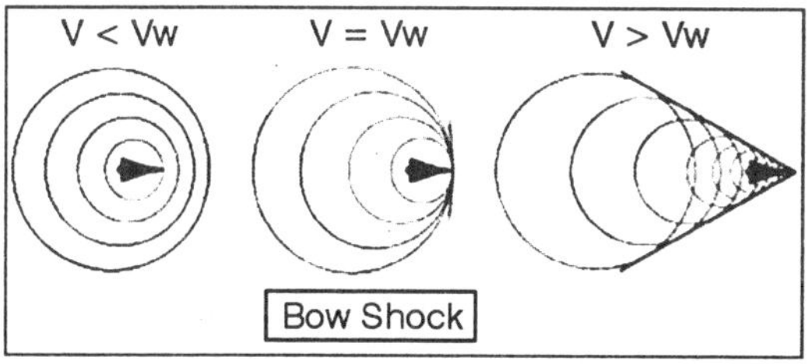

- It was mentioned earlier in this chapter that musical instruments normally generate a spectrum of sound frequencies for each musical note. The fact that waves are additive means that all these "component" sounds, consisting of nice clean sine waves, actually constitute a single, messy-looking sound waveform.

This seems puzzling at first, since it leads to the question: which is for real? The spectrum of component sounds? Or the single messy-looking sound waveform? The answer is: they both are. The tone could be interpreted in either way depending on which is more convenient to the task at hand.

A branch of applied mathematics known as "Fourier analysis" has been developed to translate between the two aspects of the sound, as well as other, similar wave phenomena. The sound can be inspected either in the "frequency domain", in the form of a spectrum of simple sine waves, or in the "time domain", in the form of a messy single waveform. Further discussion of Fourier analysis is far beyond the scope of this document.

MECHANICAL RESONANCE

- We have already discussed how sound can be created from a vibrating string or from a resonant chamber, such as a whistle. This is related to the concept of "structural resonance", or the notion that structures tend to oscillate at certain specific frequencies when subjected to forces. The most common example is that of a set of wind chimes. If one of the chimes is struck, it vibrates at a certain frequency, and this frequency increases for smaller chimes, since they can vibrate more rapidly.

Another example of mechanical resonance that was of great practical use for centuries is the "pendulum", the swinging weight on a rod that times a grandfather clock. The weight on the end of the pendulum swings back and forth, with maximum kinetic energy and speed at the bottom of its arc and maximum potential energy as it momentarily halts and reverses direction at the ends of its arc. Although a detailed analysis of pendulum motion is beyond the scope of this document, very interestingly in principle the period of a pendulum only depends on the acceleration of gravity and the length of the rod as follows:

$$\text{Period} = 2\pi\ \text{I} \times \sqrt{(\text{acceleration due to gravity/length of rod})}$$

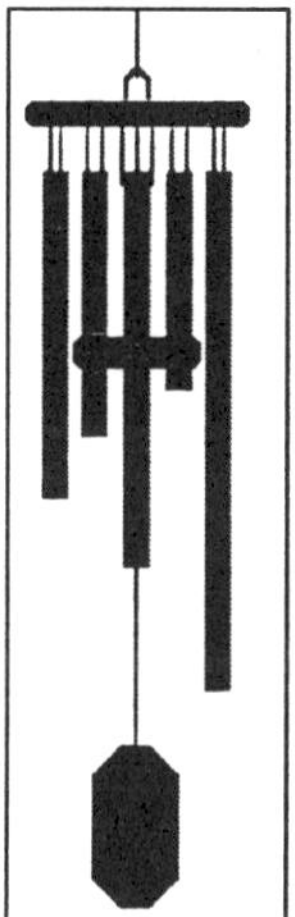

Wind Chims

This neat property, which was discovered by Galileo, allows a pendulum to count the ticks for a clock, an application that was invented by Christiaan Huygens. Since there are always some friction losses in the operation of the clock that cause it to run down, the clock includes a mechanism, either driven by a coiled "mainspring" or a weight falling on a chain, to provide a little force that keeps the pendulum in motion, while the pendulum keeps the mechanism operating at predictable rate.

The mechanism links to the pendulum through a gear wheel with jagged teeth known as an "escapement wheel"; and the pendulum links to the escapement wheel through a "pivot", a pivoting arm with two teeth that rocks back and forth in step with the pendulum. The force of the teeth of the escapement wheel against the teeth of the pivot keeps the pendulum in motion, while the back-and-forth motion of the pivot keeps the teeth of the escapement wheel stepping at a fixed rate, providing the "tick-tock" as the clock counted off the seconds.

Since the length of the pendulum rod tends to change with temperature from summer to winter, precision pendulum clocks were once made that had rods assembled from segments of dissimilar metals so their length would remain more constant. Much effort and ingenuity was put into developing highly accurate pendulum clocks.

One of the indications of their accuracy was that when clocks were moved from high latitudes to equatorial regions, their timing clearly seemed to run faster. As it turned out, the equatorial bulge of the Earth means that the acceleration of gravity is very slightly less than it is at high latitudes, resulting in a slightly but shorter pendulum period that became obvious over relatively long periods of time.

- Pendulums are very simple systems, but they illustrate several concepts in basic physics. For example, if the travel of a pendulum is charted over time, it will create a neat sine curve. Another interesting aspect of the physics of the pendulum is that it can be used to demonstrate the rotation of the Earth as a simple consequence of the First Law of Motion.

If a pendulum is made of a weight hung from a wire so that it can swing back and forth in any direction in the horizontal plane, then if the pendulum is placed at the North or South Pole, its direction of motion will remain unchanged as the Earth spins beneath it, or from an observer's point of view the direction of its swing will pivot around in a full circle every 24 hours.

If such a "Foucault pendulum" is taken to lower latitudes, the time required for it to shift its direction of swing will increase. At the equator, its direction of swing will not change at all, because then the change in the Earth's direction is in the pendulum's vertical plane, not its horizontal plane. Large Foucault pendulums are often displayed in science museums, at least at high latitudes. However, this is drifting away from the topic of mechanical resonance.

- There is another variation on the basic pendulum, called a "torsional pendulum", which involves a weight spinning at the end of a wire, twisting up the wire until the weight comes to a halt and then reverses direction. Analysis of a torsional pendulum is a little more complicated than that of a normal swinging pendulum, but the period of a torsional pendulum is also relatively constant, and it has also been used as a timekeeping element of clocks.

Since including a swinging pendulum on a small portable desktop clock or clockwork watch wasn't practical, clockmakers developed a variation on the torsional pendulum, known as the "hairspring".

This is a loosely coiled spiral of metal placed inside a relatively massive ring known as a "balance wheel"; the inner end of the hairspring is attached to a fixed axle, while the other end is connected to the interior of the balance wheel. The balance wheel is set to spinning one way, tightening up the hairspring, and then the hairspring unwinds to send the balance wheel in the other direction, with the cycle repeating over and over at a constant rate. The period of the hairspring could be "tuned" with a moveable arm that limited the maximum radius of its extension.

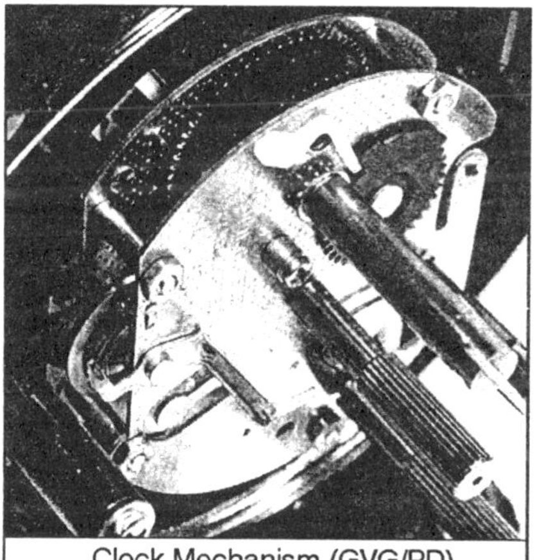

Clock Mechanism (GVG/PD)

For a desktop clock, the position of the hairspring wasn't critical, but for watches, which move around a good deal, it was placed at the centre of the watch to make sure the external forces on it were as neutral as possible to keep from affecting the assembly's period. As with the swinging pendulum, both the torsional pendulum and the hairspring assembly interface to the rest of the clock through a pivot and escapement wheel.

- Yet another variation on the pendulum is a bobbing pendulum, consisting of a weight attached to a spring, bobbing up and down. Since the motion of this pendulum tends to run down quickly, this scheme is not used much or at all in clocks or other machines, but the mathematical analysis of such a system is straightforward, and it is used as a tool in teaching the theory behind mechanical resonance.

Since an idealized spring has no losses, a third element called a "dashpot" is added to incorporate losses into the system. Analysis of such a "spring-mass-dashpot" system is yet another thing beyond the scope of this document, though it will be quickly encountered by readers who go on to college-level basic physics or engineering courses.

- Machines or buildings can have unintended and unwanted structural resonances. One of the items in the test programme for products such as computers or power tools is to put them on a "vibration table" and see how well they stand up to vibration, and particularly to see if there are certain resonant frequencies where they vibrate all too well and shake themselves to pieces. One of the elements in aircraft flight test programs is to identify flight conditions in which the wings or other elements of the machine will oscillate, or "flutter", possibly leading to loss of control or damage to the aircraft.

Skyscrapers tend to sway slightly in high winds, acting as inverted pendulums fixed to the ground, and will oscillate back and forth with a resonant period of several seconds. Under certain wind conditions, such oscillation could be a threat to the building, since the motions could build up, like a kid "pumping up" a playground swing, until they bring the house down.

The most famous example of destructive structural resonance occurred on 7 November 1940. The Tacoma Narrows Bridge, a new 850 meter long, slender suspension bridge in the northwest US state of Washington, proved to oscillate so badly in crosswinds that it was nicknamed "Galloping Gertie". When the structure was hit with a steady 68 KPH wind on that day, it began to oscillate at its resonant frequency, the wind twisting the roadbed one way and then the other as it turned up and down, and collapsed a little over an hour later. The only casualty was a dog, but in the aftermath of the accident some other suspension bridges were reinforced to damp out such oscillations. The bridge was rebuilt a decade later in a more robust form.

The Citicorp skyscraper in New York City, mentioned in an earlier chapter, incorporates an "oscillation damper" at its top to reduce its sway and damp out oscillation. The damper consists of a 365 tonne concrete slab that is moved back and forth under control of motion sensors attached to the building's structure.

If the building begins to sway one way, the slab is shifted in the other direction to compensate. Implementing the damper was cheaper than adding wind bracing to the building's structure.

It is also possible in principle to build dampers using water tanks, with large tanks at the four sides of top of the building and cross-connected by pipes at the bottoms of the tanks. Under normal circumstances, the water level will be the same in all tanks, but if the building sways, inertia will cause the water to slosh through the pipes to oppose the sway.

The tank system is designed so that its own rcsonant frequency is the same as that of the building, with the antiphase resonances of the building and the tank system tending to cancel each other out.

Chapter 11

Light and Optics

The concept of mechanical waves leads to the concept of light, which is also a wave phenomenon. However, its nature is somewhat more devious than sound waves and other mechanical waves, and in fact has some maddeningly contradictory aspects that eventually broke through the boundaries of classical physics. This chapter provides a short description of the behaviour of light.

THE DISCOVERY OF LIGHT / ELECTROMAGNETIC RADIATION

- The concept of light is as fundamental as that of matter and predates recorded history. Obviously there was something present during the daytime that allowed people to see that wasn't around at night, and that light was emitted by hot objects such as the Sun, or by fires. It was also obvious that light carried energy, since it brought warmth; that it came in the different colurs of the rainbow; and that it could be reflected by shiny surfaces, or refracted by passing through water or crystals.

The ancient Greeks had some confused ideas about light, such as failing to understand that the eye operated by passively receiving the light that fell on it, instead believing the eye actually generated the light. They still got some things right. The philosopher Empedocles (~490:430 BC) speculated that light had a finite speed, and Aristotle correctly linked the occurrence of rainbows to the reflection of light from raindrops, though he had no real idea of why the result was a colured band. The mathematician Euclid (around 300 BC) correctly understood the rules of reflection of mirrors.

The confusion over the operation of the eye was finally resolved through experiments performed by the Egyptian scientist Ibn Al Haythen, known in Europe as "Alhazen" (~965:1038). However, the modern understanding of optics did not begin until the 17th century, as researchers began to determine the precise rules of reflection and refraction, leading to the introduction of optical equipment, such as eyeglasses, microscopes, telescopes, and cameras.

- Following the beginnings of modern optics, scientists debated for over a century whether light was a particle or wave phenomenon. The particle theory had the advantage of not requiring any medium for the propagation of light, since there didn't seem to be any medium there to carry it, but the wave theory was far superior in explaining phenomena such as refraction. In the 17th century, this debate was carried on by Isaac Newton, who advocated the particle theory, and the Dutch scientist Christiaan Huygens, who advocated the wave theory.

In 1803, Thomas Young the inventor of Young's modulus published studies of the behaviour of a light source when passed through two narrow slits. If light were a particulate phenomena, the light falling on a surface beyond the slits should have been concentrated directly behind the slits, but what he observed was a pattern of alternating light and dark lines, the product of wave interference. In hindsight, that was enough to tip the scales toward the wave theory of light, but such was the authority of Newton that it took another 20 years for people to come around.

The clincher to the wave theory was provided by the French physicist Augustin Jean Fresnel, who performed an analysis that showed that the shadow cast by a small solid sphere in a bright light should have a bright centre. This was due to the fact that light diffracted around the edges of the sphere, which wouldn't have happened if light had been a particle. Fresnel performed this analysis to a board of inquiry that didn't want to believe it, but the board found his evidence irrefutable. That seemed to resolve the debate in favour of the wave theory, though it still left open the problem of the medium that light used for propagation.

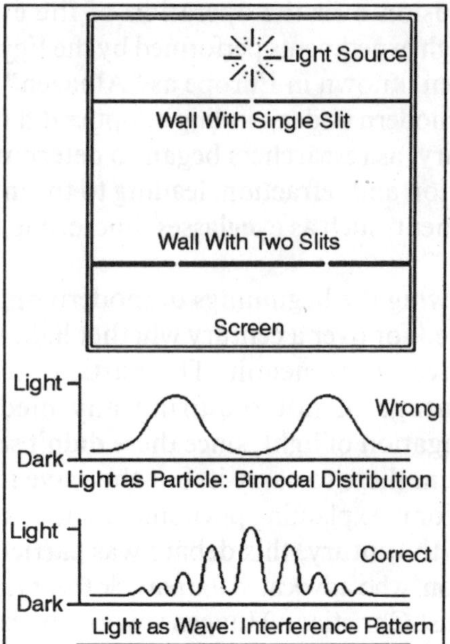

Two-Slit Interference Experiment

Of course, Newton was perfectly aware of light interference effects, having devised an experiment known as "Newton's rings". This involves placing a glass lens, curved side down, on a mirror. Light passing through the lens will be partly reflected at its bottom face, with the rest reflected from the mirror, with the two components interfering to create a set of concentric rings.

Newton knew this was a wave phenomenon, but suggested that it was due to some wave set up by the passage of light particles through the glass. In terms of 20th-century physics, Newton was in some ways on the right track to the truth, but that is another story.

- It is very easy to observe light interference at home. Simply hold up your hand with your fingers pressed together and observe a bright light through the cracks between the fingers. You will see one or more fine black hairlines in the crack, which are due to light interfering as it passes around the edges of your fingers.

Optical interference effects have practical uses. They are the basis of "interferometers", highly sensitive instruments in which two light beams sent through alternate paths are allowed to interfere, with any slight change in the relative paths of the two beams causing alternations in the intensity of light output.

An interferometer can be used as an "optical gyroscope" by arranging the two light paths as alternate routes from one vertex of a square to the opposite vertex, with the rotation of square around its centre causing the effective path of light along one side to shorten and around the other side to lengthen. Another scheme is to fit some class of sensor into one of the paths, with the sensor's optical properties changing with the parameter to be measured.

Interference effects are also used in "holography", with interference patterns from the reflections of light off an object recorded in photographic film as a "hologram". Shining light through the hologram will reconstruct the image in three dimensions, and the image will seem to float in space.

- A half century after Thomas Young, James Clerk Maxwell conducted a brilliant theoretical analysis of electric and magnetic fields, concluding, among other things, that these fields could propagate through space in a coupled fashion as a wave.

  Maxwell's analysis not only identified light as such an "electromagnetic wave", but also predicted that such electromagnetic waves would exist over a very wide range of frequencies, most of which were invisible to the eye.

Scientists had known for decades there were invisible forms of light. Early in the 19th century, the Anglo-German astronomer Sir William Herschel used a prism to break sunlight up into its rainbow spectrum, and then moved a thermometer through the spectrum to see how much warmth it contained. The temperature rose steadily from the blue end of the spectrum to the red end of the spectrum, and then continued to rise as Herschel moved the thermometer past the red end of the spectrum. This invisible form of light was named "infrared" light.

Maxwell's analysis extended this concept by showing that electromagnetic radiation existed over an infinite range of frequencies, and that only a limited range of these wavelengths were visible to the eye. Visible light quickly became only a small subset of the entire "electromagnetic spectrum". Long wavelength electromagnetic radiation became identified as "radio waves" that could be generated and detected with electronic gear, leading to the foundation of radio communications.

- One of the questions that arose with early investigations of light was whether it had a finite speed. Galileo and an assistant performed some experiments, shining lanterns at each other from distant hills, but all he was able to conclude from such a clumsy scheme was that if light had a finite speed, it was very fast.

In 1676, the Danish astronomer Olaus Roemer noticed that the predictable timing of the eclipses of the moons of Jupiter shifted relative to the position of the Earth, advancing when the Earth was on the same side of the Sun as Jupiter and falling back when the two planets were on opposite sides of the Sun. He was able to obtain an estimate of the speed of light by measuring the delay and using estimates of the size of Earth's and Jupiter's orbits; his estimate was that light traveled at about 227,000 kilometers per second. That was about 75 per cent of the real value not on the mark but well in the ballpark, and definitely on the right track.

In the 19th century, improved technologies permitted a much more accurate measurement of the speed of light. The French experimentalist Armand Hippolyte Louis Fizeau used a system of rotating toothed wheels, intercepting a light beam that was routed between them over a roundabout path with a set of mirrors to increase the path length.

The speed of rotation of the wheels was increased until the light was blocked, allowing calculation of the speed of light from the path length, the size of the gap between the teeth of the wheels, and the rate of rotation. His colleague Jean Bernard Leon Foucault, best known for his invention of the Foucault pendulum, performed even better measurements using a system that was based on spinning mirror drums instead of toothed disks. Foucault's value turned out to be very close to 300,000 kilometers per second.

THE ELECTROMAGNETIC SPECTRUM

- Electromagnetic (EM) radiation propagates through a vacuum as a wave moving at (very close to) 300,000,000 meters per second. This gives the relationship between frequency and wavelength:

$$\text{wavelength} = \frac{300{,}000{,}000}{\text{frequency}}$$

EM radiation occurs over a continuous spectrum of wavelengths, ranging from low-energy long wavelengths at the bottom of the spectrum to high-energy short wavelengths at the top of the spectrum. The visible light that humans can sense is only a very small part of the full EM spectrum. The spectrum extends through longer wavelengths, down through the "infrared", "microwave", and "radio" regions; and through shorter wavelengths, including the "ultraviolet", "X-ray", and "gamma-ray" regions.

The Em Spectrum

Radio	30 centimeters or longer
Microwave	1 millimeter to 30 centimeters
Infrared	0.77 microns to 1 millimeter
Visible	0.39 to 0.77 microns
Ultraviolet	10 nanometers to 0.39 microns
X-ray	10 picometers to 10 nanometers
gamma ray	less than 10 picometers

A "micron" is a micrometer, or millionth of a meter. There is actually overlap in the definitions of all the bands except the visible band, and the figures given here are useful simplifications The energy of EM radiation increases with frequency. Radio waves and microwaves aren't very energetic, though as a microwave oven demonstrates they can still transfer a lot of energy if produced in quantity.

High-frequency ultraviolet waves,×rays, and gamma rays are very energetic, but fortunately the ozone layer blocks most of this high-energy radiation and prevents it from doing us harm.

The visible-light region of the spectrum is divided into "colour subbands". These range from low-energy red and orange bands, to more energetic and longer-wavelength yellow and green bands, to high-energy blue and violet bands. Light emitted by the Sun and by incandescent light bulbs consists of a mix of these bands, and is known as "white light".

Colour Subbands

Red visible	0.622 to 0.770 microns
Orange visible	0.597 to 0.622 microns
Yellow visible	0.577 to 0.597 microns
Green visible	0.402 to 0.577 microns
Bl ue visible	0.455 to 0.402 microns
Violet visible	0.390 to 0.455 microns

Isaac Newton was one of the first to list the colurs of the visible light spectrum. He defined seven colurs, not six, inserting "indigo" between blue and violet. Few could ever make it out as anything more than dark blue, and apparently the only reason he used it was because he considered the number "7" as having a special mystical significance.

LIGHT EMISSION

- Light may be emitted by matter when it is energized. One of the most common ways in which light is emitted occurs when a solid, liquid, or gas is heated, a process that is known as "incandescence".

The light from an incandescent solid or liquid object is emitted in a continuous range of frequencies, or "continuous spectrum". The continuous spectrum emitted by a solid or liquid object generally approximates that produced by a "perfect emitter" or "blackbody". This is generally modeled as a hollow sphere with a hole in it, with the radiation produced in the interior of the sphere observed through the hole. For this reason, blackbody radiation is also sometimes called "cavity radiation".

In a blackbody, the intensity of the light across the frequency spectrum rises to a peak corresponding to a specific temperature, and then declines rapidly for higher frequencies. As a blackbody is heated and grows hotter, it first emits in the infrared, then glows dull red, next turning orange as it gets hotter, then yellow, then white, and finally bluish-white, with the colurs corresponding to increasing peak frequencies. As the blackbody is heated even more, emission begins to occur in the ultraviolet, then in the X-ray region and finally the gamma-ray region of the spectrum.

The blackbody would still appear to be bluish-white to the

eye, though anyone close enough to inspect it once it got that hot would be much safer someplace else. The total energy output of the blackbody increases at approximately the fourth power to temperature, meaning that if the temperature is doubled, the energy output is increased by sixteen (2^4) times.

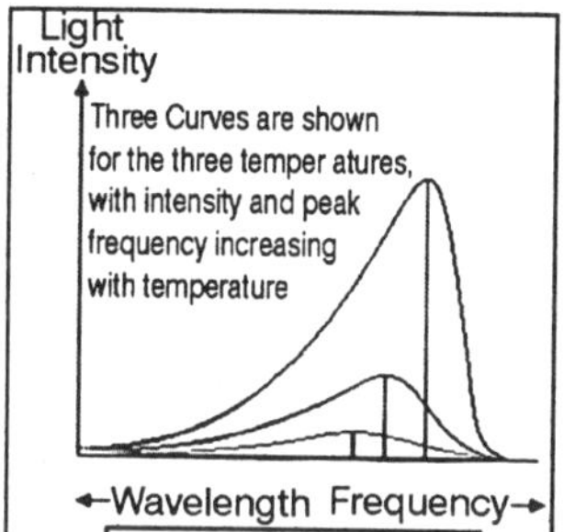

Blackbody Radiation

Stars are reasonable if not perfect approximations of blackbody emitters. Relatively cool stars are red, while very hot stars are bluish white. Small cool red stars have relatively low energy output and live for a very long time, while large hot stars use up their energy rapidly and so have short lives. A more mundane example of thermal emission is the ordinary incandescent light bulb, which glows by heating a filament with an electric current.

Most of the light emission from objects in our surroundings is in the infrared. This emission allows infrared scopes and cameras to detect objects in the dark. The appearance of objects seen from their thermal emission can be confusing since it is very different from that seen in daylight, and FLIR imagers often have a switch to flip the scene from "white hot" to "black hot" to help a user figure out a scene.

- Incandescence in gas works somewhat differently than it does in solids and liquids. Changes in the energy state of electrons in gaseous atoms and molecules cause a hot gas to emit electromagnetic radiation only at certain frequencies that are specific to the gas. This pattern is known as the "line emission spectrum" of the gas. Gas-filled lamps used in front-window displays of bars and other commercial establishments are filled with gases excited by an electric current, with the colour determined by the type of gas. Neon, for example, gives an orange light.

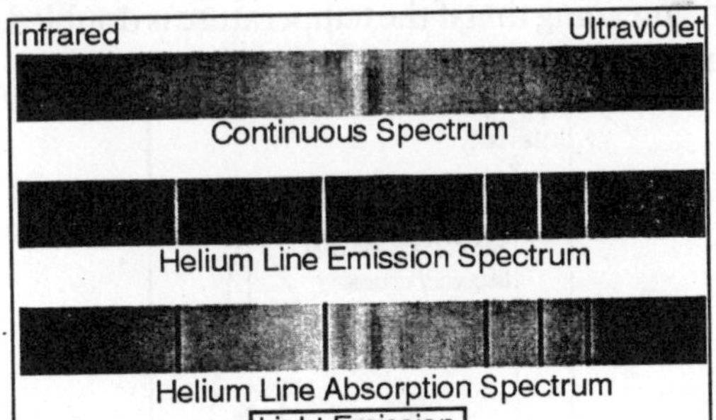

Light Emission

Interestingly, if white light is shined through a cool gas, the gas will *absorb* light at these specific frequencies, forming a "line absorption spectrum" that consists of dark lines across the rainbow spectrum in the background. The specific pattern of lines in an emission or absorption spectrum can be used to determine the chemical composition of everything from atmospheric gases to distant stars. This technique is known as "spectroscopy".

The basic concept was proposed in 1823 by Sir John Herschel, the son of Sir William Herschel, and the details were fleshed out by two German researchers, Robert Wilhelm Bunsen, and Gustav Robert Kirchhoff. They collaborated to develop the first formal "spectroscope" to observe line emission spectra and also characterized the line spectra of many gases. The basic rules of line spectroscopy, as described above, are known as "Kirchhoff's laws". Such line spectra can also be used to determine the velocity of an object along our line of sight (known as its "radial velocity"). As with sound waves, light can be Doppler-shifted, and the entire line spectrum of a cosmic object will shift in frequency by an amount proportional to its radial velocity. If an object is coming towards us, the frequencies will be higher, or "blueshifted"; and if it is going away from us, the frequencies will be lower, or "redshifted".

- There are also "luminescent" light emission processes. The best known is the "arc discharge" due to a lightning bolt, which ionizes the air and causes the electrons to emit light. For another example, if some substances are exposed to high energy light, they may emit light at a lower energy, a process known as "fluorescence".

In a fluorescent light tube, an electric current flowing through a mercury gas produces a wide range of wavelengths, which then excite a "phosphor" material coating the inside of the tube to emit visible light. Some of these substances will continue to emit light in a fading fashion after the energy source is turned off, a phenomenon known as "phosphorescence".

Chemical and biochemical processes can emit light as well. One of the more interesting modern examples is the "light stick" toy, which when twisted mixes two chemicals that react and emit pale light, a process known as "chemiluminescence". Many biological organisms also emit light by chemical processes, though the emission is called "bioluminesence" in this case.

Fireflies are the best-known example, though sailors who have passed through tropical seas are also well aware of marine microorganisms that will emit light when disturbed, leaving a glowing trail as the bow of a ship plows through the water.

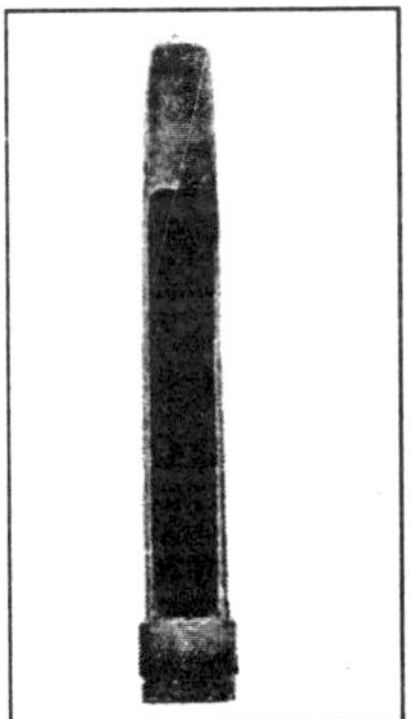

Lightstick (GVG/PD)

There are "triboluminescent" processes that involve light emission due to stresses on a material. This can be easily demonstrated by quickly twisting a plastic tray of icecubes in a dark room: It will emit a faint flash of pale blue light. Intense sound waves can also create light emission in certain materials, a process known as "sonoluminescence".

- Modern lasers involve light-emission principles that would have not been entirely familiar to classical physicists.

  A gas laser is basically a gas-filled lamp built as an optical resonant cavity, with a mirror at one end and a partly-transparent mirror at the output end and the length of the cavity being some integer multiple of one of the wavelengths of line emission for the gas. Energizing the gas-filled lamp causes light to build up in the resonant cavity and leak out the partly transparent mirror as a tightly focused beam. Classical physics would have found the fact that the light in the beam is all in phase "coherent" difficult to explain.

Some classes of solid-state lasers substitute a fluorescent material such as a rod of ruby crystal for a gas-filled tube; not incidentally, fluorescence is a line-emission process. However, the solid-state lasers like those used in CD or DVD players are a concept that would seem like to complete magic to a classical physicist and so have to be discussed elsewhere.

BEHAVIOUR OF LIGHT

When light strikes matter, it can be absorbed, reflected, or passed through. A black plush cloth will absorb most of the light; a mirror will reflect most of the light; and a sheet of clear glass will allow most of the light to pass through.

Reflection is a simple process. If a beam of light strikes a smooth, flat reflective surface such as a mirror at an angle, it will bounce off the mirror at the same angle as it struck. Of course, if the surface is rough, it will be scattered instead of neatly reflected.

The angle of reflection does not change with the wavelength of the light. Light passing through a transparent material tends to be refracted, with the amount of bending depending on the material's "index of refraction", which is given by:

$$\text{Index of refraction} = \frac{\text{speed\_of\_light\_in\_vacuum}}{\text{speed\_of\_light\_in\_material}}$$

The index of refraction is close to 1 in air, about 1.3 in water, about 1.5 in glass, and about 2.4 in diamond. Refraction can be seen by observing from certain angles a spoon stuck in a glass of water, which appears bent under the water.

A precise index of refraction can in general be only given for a single wavelength of light, since it varies with wavelength. This is why a prism creates a tiny "rainbow" when sunlight shines through it, since the different wavelengths of light bend in a range of different directions. This process is known as "chromatic dispersion". The amount of bending for a material with a given refractive index is given by "Snell's Law", after the Dutch mathematician Willebrord Snellius or Snell, which is very important for the design of lenses for optical instruments.

If a beam of light is passing through a medium with refractive index R1 and strikes the surface of a medium with a refractive index R2, then the angle of the beam relative to the vertical in the two mediums is given by:

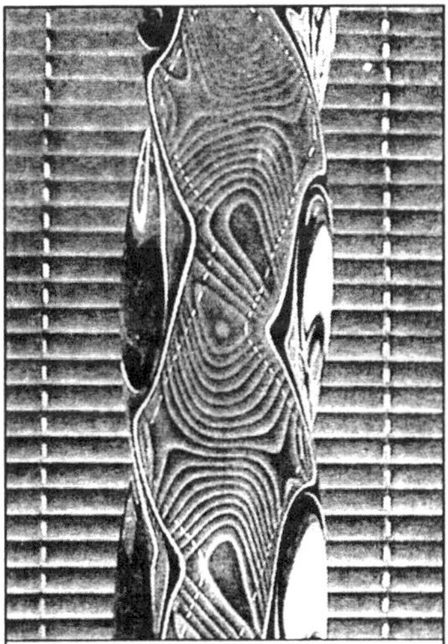

REFRACTION EFFECTS (GVG / PD)

$$R_1 / R_2 = SIN(angle_1) / SIN(angle_2)$$

where SIN gives the trigonometric sine function.

Reflection and refraction are the cause of rainbows. If Dexter has the Sun to the back of his head and observes a rainstorm, light from the Sun passing through the rain droplets is partially reflected and bounced back towards him, experiencing refraction and chromatic dispersion as it passes through the droplet.

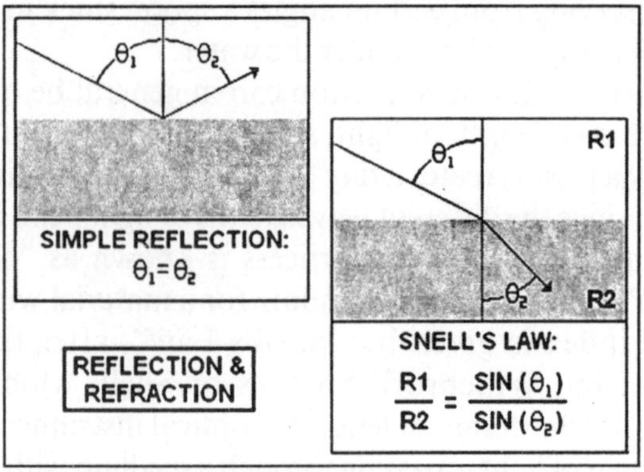

The rainbow will appear at an angle of about 40 degrees around the line established by the Sun and Dexter's head, though of course the angle varies a bit with colour.

DOUBLE RAINBOW (GVG / PD)

Remnants of the sunlight will bounce around through the droplet again, and if the sunlight is bright enough and the rain dense enough, Dexter will see a faint "double rainbow", at about 50 degrees from the line from the Sun through his head. It is sometimes even possible to see a short and faint section of a triple rainbow, but it is very rare.

Another unusual occurrence, though not quite as rare, is a "halo" that appears around the Sun on very humid and hazy days, which is caused by sunlight passing directly through water droplets and being refracted into its spectral components. Refraction is the cause of desert mirages. Cold air has a higher index of refraction than warm air, and the air close to the hot desert sands is hotter than the air above it. The temperature gradient through the air causes light to curve upward, with the result that distant objects seem to "float" upward, which can cause mis-estimations of distances and other confusion.

Incidentally, if light strikes a medium of different refractive index at a shallow enough angle, all the light will be reflected. This phenomenon is known as "total internal reflection". One application of total internal reflection is "optical fibers", the threads of glass sometimes used in novelties and for optical data-communications links. These optical fibers have a glass core with one index of refraction and an outer layer or "cladding" of another, causing light sent down the axis of the fibre to be reflected down its length with few losses.

As mentioned earlier, light is also diffracted, bending around the edges of objects. Diffraction, like refraction, is wavelength dependent, and can produce rainbows from a light source. Astronomers performing spectroscopy of stars and other cosmic objects normally do not use a prism to obtain the line spectra of those objects, since a prism would have a narrow field of view.

Instead, they use a "diffraction grating", which is essentially a grooved mirror surface, with the reflected light diffracting around the edges of the grooves.

A good home-grown equivalent of a diffraction grating is a CD or DVD disc: if held to the light it displays pretty colurs, due to diffraction from the disc tracks.

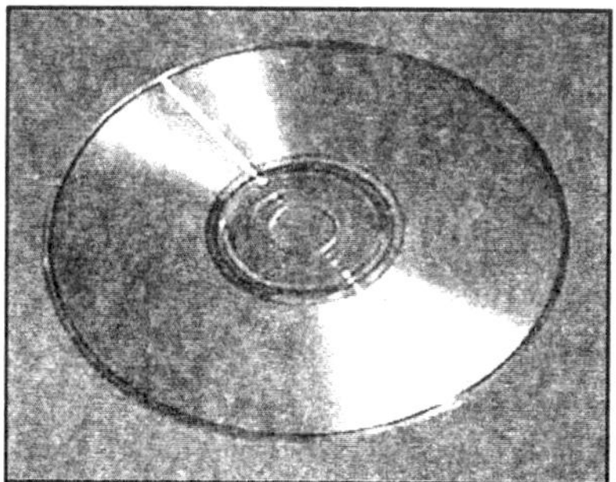

DIFFRACTION EFFECTS (GVG / PD)

- The actual colurs we perceive in the real world arise from a number of processes. Thermal emitters like the Sun and line emitters such as neon lamps produce their light directly, but all other objects produce their light by reflection.

Objects with light-colured surfaces reflect most of the light falling on them, while those with dark-colured surfaces absorb most of the radiation falling on them, with the energy then released as thermal emission.

Some objects may selectively absorb some wavelengths and reflect others. Plants absorb blue and red wavelengths to perform photosynthesis, while reflecting green wavelengths, which is why most plants appear green.

The reflectivity of an object is also dependent on its smoothness. A very smooth surface reflects light without much scattering, and is known as a "specular" or "mirror" reflector. Calm bodies of water are specular reflectors. A rough surface scatters light, and is known as a "diffuse" reflector.

Except for the few of us who have been astronauts, we observe light that has passed through the atmosphere. The atmosphere not only blocks light through hazes and clouds, but also selectively scatters light at the shorter wavelengths.

Scattering from the molecules in the atmosphere diffuses short-wavelength blue light from the Sun, causing it to appear all over the sky and so making the sky blue.

On the other side of the coin, this same scattering of blue light makes sunsets tend to be red: the shorter "bluish" wavelengths of light are scattered, subtracting them from the Sun's light and leaving longer "reddish" wavelengths behind. Smoke from forest fires or heavy industrial pollution dumps large sooty particulates into the air, which cause even greater reddening, resulting in dramatic "blood red" sunsets.

Incidentally, it is only possible to see a light beam from the side because of scattering. The well-known pyramidal Luxor Hotel in Las Vegas has searchlights in its apex shining up into the sky, which the designers believed would cap the building with a spectacular column of light.

The effect wasn't as spectacular as hoped, however, since in the clear desert air the beam was generally invisible to passer-bys, except when a dust storm was in progress, when nobody wanted to be outside to enjoy the view.

BLUE SKIES (GVG / PD)

- Light is a transverse wave. This means that is it is "polarized", with an angle of oscillation ranging 360 degrees around its direction of propagation. A "filter" can be designed to pass light of one polarization, but not another. For example, the lenses of "Polaroid" sunglasses contain long-chain molecules that are all oriented in a straight line in one direction.

  Light can pass through the lens if the polarization is in the same direction as that of the molecules, but not at other polarization angles. Reflections from horizontal surfaces tend to produce horizontally polarized light, and Polaroid sunglasses use vertical polarization filters to get rid of this glare.

It is a fun little game to take two sets of Polaroid sunglasses and look through the lenses of them in series at right angles to each other; the light is blocked out. (However, as will be described later, there is a very devious bit of fine print to this experiment.)

THE EYE AND COLOUR

- Of course, we sense light through our eyes. The human eye has two classes of sensors, named "rods" and "cones". Rods provide perception of large patterns in the field of view, while cones provide detail and colour perception. There are three types of cones, with each sensitive to a different region of the visible-light spectrum:

- "Blue" cones are most sensitive to short wavelengths of visible light, centered around a wavelength of 0.44 microns.
- "Green" cones have a peak sensitivity around a wavelength of 0.53 microns.
- "Red" cones have a peak sensitivity around a wavelength of 0.57 microns, which is actually in the yellow band of the visible spectrum. However, the red cones still have strong response at redder wavelengths.

Incidentally, this arrangement was discovered by Thomas Young, whose formal training was as a medical doctor. Young was one of the truly brilliant figures in the history of science, with his accomplishments even including the deciphering of Egyptian hieroglyphics. However, he was the sort of genius who could not communicate with others on their own level, and he was a notoriously tiresome lecturer.

The perception of colurs depends on the ratio of stimulation of these three types of cones by light, or in other words on the ratio of the red, green, and blue (RGB) components. These three "primary colurs" add as follows:

red + green + blue -> white
red + green -> yellow
red + blue -> magenta
green + blue -> cyan

Different shades are produced by varying the level of each of the red, green, and blue components. Such an "additive" scheme is used to generate shades of colour on a colour TV or a computer display.

The screen of a colour picture tube, for example, consists of a display with triplets of precisely-arranged dots of red, green, and blue "phosphors" that glow when hit by an electron beam used to "write" the picture on the screen, scanning in rows from the top to the bottom of the screen. The tube actually has three electron beams that scan in parallel, each painting a red, green, or blue image that add up to a full-colour image.

Some tubes use vertical stripes of phosphors, but the effect is the same. Incidentally, there was an early colour TV scheme that actually generated the red, green, and blue images sequentially one after the other, scanning them out rapidly enough so that it came out as a full-colour image. The three images were all generated in black and white on an internal screen and then projected to a viewing screen through a rotating RGB colour wheel. The scheme was simple and effective, but such a TV did not work with the black and white TV broadcast signals that were then the standard, and it was not successful.

The scheme that did become established amounted to a standard TV signal that provided a black and white image, with an added signal component that effectively "painted" colour on to the black and white image. A black and white TV set simply ignored the colour component.

Astronauts on the later Apollo Moon landings did carry a sequential-colour TV camera, since its simplicity allowed it to be made much smaller and lighter than the conventional parallel colour TV cameras then in use. This sequential camera was designed specially for the Apollo project, and the sequential images were converted to standard colour signals after being radioed back to Earth.

- Somewhat confusingly, printed colour images use a complementary "subtractive" colour scheme, based on yellow, magenta, and cyan dyes.

yellow	+ magenta	+ cyan	->	black
	magenta	+ cyan	->	blue
yellow		+ cyan	->	green
yellow	+ magenta		->	red

The subtractive scheme is based on the fact that pigments are passive reflectors of ambient light, which is generally close to white light. Magenta pigment, for example, reflects blue and red light but absorbs green.

Since yellow pigment reflects red and green light but absorbs blue, then a mix of yellow and magenta absorbs both blue and red light, while reflecting green. Mix yellow, magenta, and cyan paint together, and it absorbs all colors, leaving only black.

Perceived colors are also dependent on brightness, since rods tend to dominate over cones as light levels drop. The eye is very sensitive to colors, able to distinguish hundreds of thousands of different colors.

However, the eye's ability to automatically compensate for

changes in brightness makes it very poor at discriminating light levels, with tests showing the number of levels it can discriminate as no more than about 30.

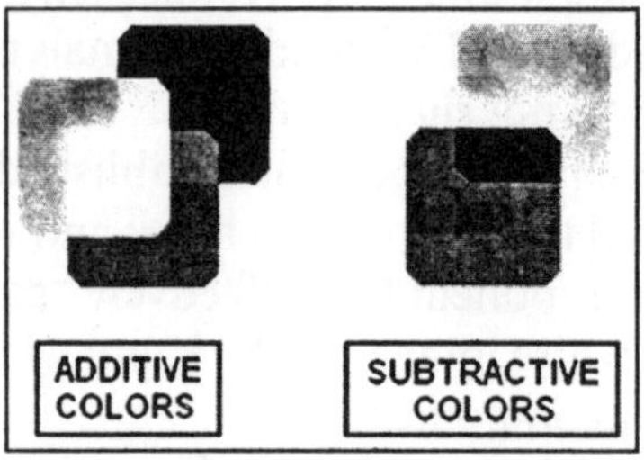

The eye's structure and the layers of nerve cells that support that structure are optimized for certain tasks, such as determining the edges of objects and their motion. The eye, interestingly, is not very sensitive to changes in shade across a scene. A large square that varies gradually in shade from side to side will appear as one colour.

- It has to be emphasized that there is nothing fundamental, at least from the point of view of pure physics, about the colors red-green-blue (or yellow-magenta-cyan). They are just arbitrary bands of light. The only reason they are special in any way is because the trial-and-error motions of evolution just happened to build cones that were sensitive to these three wavelength bands.

  Different classes of other animals with colour vision have different colour reception schemes; instead of three different types of colour receptors, most mammals only have two, while birds often have four and can see into the ultraviolet. Nocturnal anim als which are active at night and sleep during the day generally don't have colour vision at all.

Chapter 12

Optical Technology

- A knowledge of the behaviour of light has led to the development of technologies to correct eyesight, using "eyeglasses"; provide close-ups of the very small, using a "microscope"; observe the very distant, using a "telescope"; and manipulate light and images in a wide range of other ways. This chapter provides a survey of modern optical technology.

PINHOLE CAMERAS, LENSES, THE EYE, AND EYEGLASSES

- The basis of optical technologies is the bending of light rays. Any object that is generating or reflecting light can be thought of as a source of an infinite set of light rays; each point or element of the object generates its own infinite set of rays in all directions from which it can be seen, with a single ray linking the element to a particular viewer.

  This sounds exotic, but it's simple: all it's saying is that an element of an object can be viewed from any angle, with a simple straight line path that light can follow between the element and the viewer for each angle.

The best way to see this is to understand the simplest optical device, the "pinhole camera" or "camera obscura". Suppose Dexter builds a big cube that light can't get into, so he's completely in the dark when he crawls inside. However, he builds a box into one wall of the cube, with a pinhole on the front of the box to let light in, and a translucent screen on the rear of the box so he can see the light falling in through the pinhole. Somewhat surprisingly, the screen displays an upside-down view of the world as seen beyond the pinhole.

Consider a simple visual target, consisting of four different-colored diamonds clustered side by side, set up beyond the pinhole, and for simplicity consider only the rays of light coming from the precise centre of each diamond. Although there are rays of light running from the centre of the four diamonds in all directions from which the diamonds are visible, there's only one ray from the centre of each diamond to the pinhole.

These four rays pass through the pinhole and fall on the translucent screen at the back of the pinhole camera; of course the ray from the bottom diamond falls on the top of the screen, the ray from the top diamond falls on the bottom, the ray from the left diamond falls on the right, and the ray from the right diamond falls on the left. The sum of the rays from the diamonds create an upside-down image of the target on the screen.

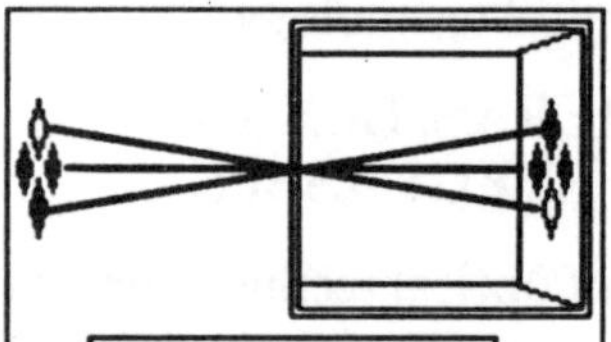

A Pinhole Camera

The image of the target will get bigger if Dexter moves the screen back from the pinhole, but that image will also get fuzzier and in particular fainter. That pinhole can only admit a small amount of light, and the bigger the image, the larger the area that the light is spread over.

Making the pinhole bigger will result in a brighter image, but also a fuzzier one, since the larger pinhole can be considered as like a set of adjoined small pinholes that are slightly offset from each other, resulting in a matching set of superimposed images that are slightly offset from each other as well. A pinhole camera is an interesting toy, but it's not generally all that useful.

- To build more interesting optical devices, Dexter needs to use lenses or mirrors. He starts out with lenses, coming up with a "magnifying glass", one of the oldest useful optical technologies. It is simply a "convex" lens, a disk of glass thick in the middle and thin at the edges, with the curvature of the lens matching the curve of a shallow parabola.

The magnifying power of a single convex lens is related to its "focal length". Suppose rays of light are sent through a convex lens directly face-on to the lens, entering the front of the lens in parallel and then emerging from the back bent at angles that converge on a single location behind the lens, the "focal point". If Dexter wants to use a magnifying glass to start a fire from sunlight, he moves the glass until the focal point sits on the item to be ignited. This concentrates all the rays of light passing through the lens onto a single point, providing enough focused energy to start a fire.

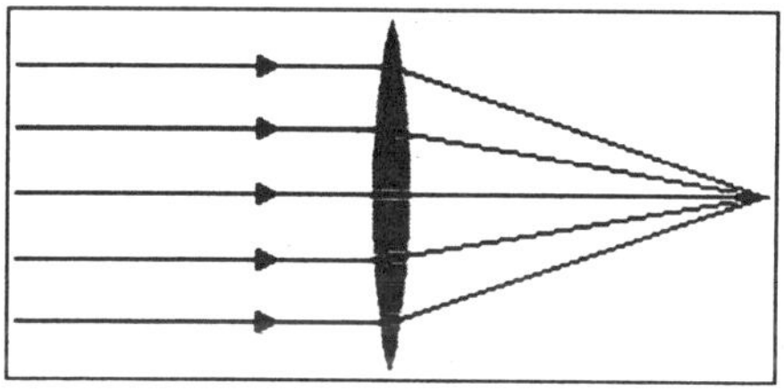

The ratio of the diameter of the lens versus the distance to the focal point from the back of the lens is the focal length. The focal length decreases as the curvature of the lens and the refractive index of the lens material increase. A parabolic curvature is required to bring rays entering the lens to a single focus; if the lens curvature were a section of a sphere instead, rays from the outer edges of the lens would come together at a closer focus than rays near the centre, a defect which is known as "spherical aberration".

Even if the lens is properly ground, since the index of refraction is wavelength dependent, light of different colors will come to different focal points, with red light coming to a different focus than blue light. This error is called "chromatic aberration".

- The magnifying power of a single convex lens increases as the focal length of the lens decreases, by the simple fact that this increases the viewing angle of the target to the eye. Let's go back to the simple visual target, consisting of four colored diamonds.

If Dexter holds it up to his eye, from a certain distance the target has a certain angular size in degrees, with the rays of light from that target covering that range of angles. If he moves the target closer, its angular size increases; if he moves it away, its angular size becomes smaller.

Now he places a simple convex lens between the eye and the target. With the lens in the way, the shortest optical paths to the four diamonds are no longer simple straight lines. The lens acts to bend or reroute rays that wouldn't have reached the eye in direct vision, giving a modified view of the target. A shorter focal length results in a greater "leverage" of the image size, resulting in an image with a larger angular size.

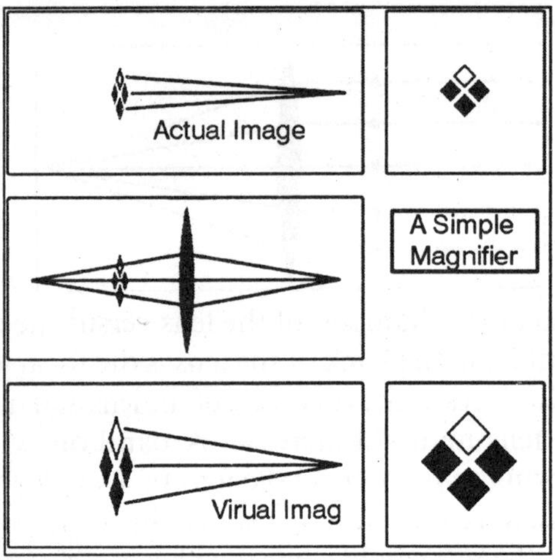

Lenses can also be built in a "concave" format, thick at the edges and thin in the middle. Rays passing through a concave lens will diverge, not converge, as they pass through the lens; this will result in *smaller* image. In addition, there is no reason why a lens can't be built in a "convex-concave" format convex on one side, concave on the other, with the light rays converging on the convex side and diverging on the concave side. The usefulness of these two lens configurations is less obvious, but they will come up later.

- The human eye is a classic example of a single-lens "device", of course well predating any human-built optical technology such as magnifying glasses. The eye is in fact so sophisticated that there are those who proclaim it as an example of "ssintelligent design" behind the natural world. It seems too elaborate to have ever been constructed by the "screening and selection of random natural changes" described by Darwinian evolution.

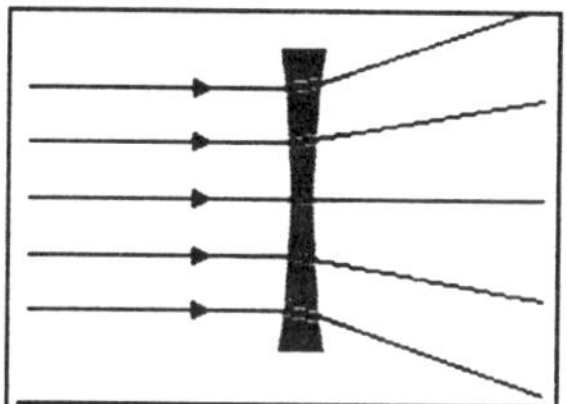

Concave (Divergent) Lens

The human eye is a fluid-filled sphere with a convex lens that focuses light on the matrix of rods and cones, discussed in the previous chapter, covering the retina on the back inner surface of the eye. The lens is protected by a transparent sheath named the "cornea" and a layer of fluid; the lens is also partly covered by a "shutter" called the "iris".

The iris opens wide in the dark to admit more light into the eye through the central "pupil", and narrows in bright light to keep the retina from being overloaded with light, or "overexposed". A set of muscles point the eye in the desired direction; another set changes the curvature of the lens to adjust its focal length to focus on near or far objects; while a third set adjusts the iris to deal with changing light conditions.

The eye in itself only receives images; also as mentioned previously, the nervous system network behind the eye and the optical interpretation sections of the brain do a great deal of "processing" on the images that gives our visual sense a number of quirks. Further discussion of the processing of images is outside of the scope of this document, but it is worthwhile to realise that our "picture" of the outside world is by no means a crisp photograph.

- It is also worthwhile to realise that not all animals have an eye like that described here. Cats and owls, which generally hunt at night, have eyes that do generally conform to this description but don't have cones, compensating with more rod cells to give more detail. Of course, insects have complicated multifaceted compound eyes very much unlike those of humans.

Darwin's defenders counter the "intelligent design" claim by pointing out the range of forms of the eye. Some microorganisms have primitive optical sensors that can hardly be called eyes, implying an evolution of capability to more sophisticated schemes; the widely varying forms of eyes, such as the eyes of insects versus those of humans, suggest an evolution through history to different structures that do much the same job; and the variation in detail of eye function between related species, such as humans versus cats, suggest an evolutionary process of adaptation of similar structures to different lifestyles.

- This is getting a bit off the subject of physics, but it does lead back to the main stream of the discussion, since another counter to the "intelligent design" claim is that the design of the eye isn't intelligent in all respects: few humans have "perfect" vision. One of the earliest applications of optical technology was a tool for correcting flawed vision: "spectacles" or "eyeglasses".

Eyeglasses are without question the most important and widespread example of optical technology, and they have been for a long time. The first glass lenses appeared in Europe in the 13th century. Not only was the new technology quickly put to use in magnifying glasses, but some bright person quickly got the idea of mounting two lenses in a wire frame that could be propped on the nose and around the ears to provide corrections for defective vision. Eyeglasses were in common use by the time of the Reformation.

Eyeglasses are designed to compensate for defects in the lens of the human eye. A nearsighted eye can only focus on close objects; nearsightedness can be corrected by concave spherical lenses, essentially bringing the images of far objects closer. A farsighted eye can only focus on far objects; farsightedness can be corrected by convex spherical lenses, essentially pushing the images of near objects away.

Eyeglasses can also be made as "bifocals", with an upper lens segment to focus on distant objects and a lower lens segment to focus on near objects. The invention of bifocals is generally credited to the great American entrepreneur, scientist, inventor, and diplomat Benjamin Franklin, who came up with the idea in 1760. Bifocals led to "trifocals", with upper, middle, and lower segments to focus on far, intermediate, or near objects. In modern times, plastic "contact lenses" that can be placed on the eye have been developed as a replacement for simple eyeglasses.

- Incidentally, although the optical technologies in this chapter generally use convex, concave, or convex-concave glass lenses, there is a wide range of alternative configurations for lenses. One of the most common is the "Fresnel lens", of course the invention of Augustin Fresnel. This is basically a short focal-length convex lens, which would normally be thick and heavy, that has been "flattened" by taking ring sections of the lens; trimming off the base of these rings; and then laying them down again concentrically in a flat plane. A Fresnel lens can't bring light to a tight focus, but it can concentrate light effectively enough, and it is used in lighthouse beacons, flashlight lenses, lenses for warning blinkers, concentrators for solar electric cells, and so on.

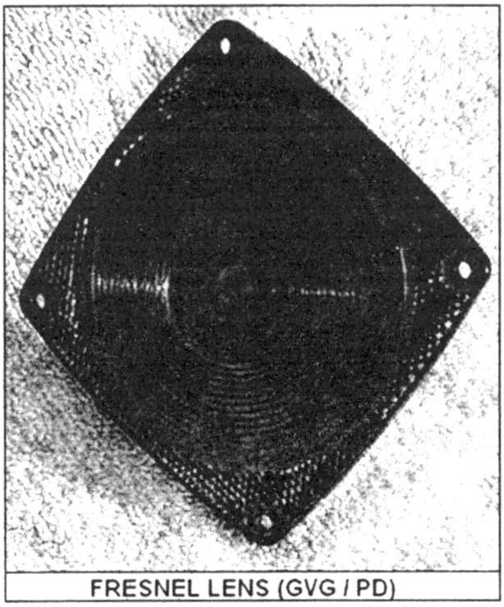

FRESNEL LENS (GVG / PD)

It is possible to build lenses based on diffractive instead of refractive principles. A "Fresnel zone plate", another one of

Fresnel's inventions, consists of an opaque disk into which concentric rings have been cut, with the rings arranged to produce the proper diffractive focus.

Fresnel zone plates are useful for focusing light in a system built on a stamped-metal chassis, saving the cost of a separate component, or in micro-sized systems where building and placing a glass or plastic lens would be difficult, In addition, it is possible to build glass or plastic lenses on diffractive principles, in the form of "binary optics", in which the lens looks like a set of increasingly smaller disks stacked on top of each other. Binary optics also works well for constructing miniaturized lenses.

MICROSCOPES, REFRACTING TELESCOPES, AND CAMERAS

- A simple magnifying glass can provide enlargements of up to about 15 times. The problem is that, since a single lens magnifies by bending light rays into a shorter and shorter focus, the focus eventually becomes too short to be usable. The solution is to use a *second* lens to perform another level of bending.

Such a "compound microscope", as opposed to the "simple microscope" or magnifying glass, uses two convex lenses mounted in a tube. The lens at the front, known as the "objective" lens, has a very steep, short focal length, giving it large magnifications of nearby objects, while the lens at the back, known as the "eyepiece", has a longer focal length, allowing it to bring the image provided by the objective lens back into a useful focus, as well as providing another factor of magnification. A compound microscope can provide magnifications of up to 2,000 times.

- Much the same concept, using an objective lens with a longer focal length mounted in a tube with an eyepiece lens, can be used to build a "telescope" to observe distant objects. Such simple "refracting telescopes" or "refractors" were introduced early in the 17th century, with Galileo being the most prominent of the early advocates.

Chromatic aberration was a real problem with refractors for about two centuries, since it created a colored halo around objects.

Telescope builders finally realized that chromatic aberration could be minimized by using a lens that was convex on both sides and made of a glass with a high index of refraction, mated to a lens that was concave on one side and convex on the other and made of a glass with a much lower index of refraction. The action of two lenses cancelled out the chromatic aberration.

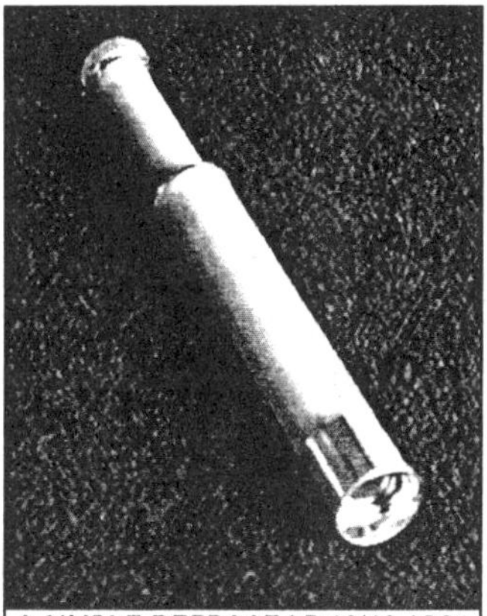

A SIMPLE REFRACTOR (GVG / PD)

A common misconception about astronomical telescopes is that their main function is to provide magnification of distant objects. This is not really the case. The interference of the continuous rippling optical variations in the atmosphere means that a really big telescope here on Earth can't get any more of a detailed image of a celestial object than a moderate-sized telescope can. This is known as the "seeing" problem; the "seeing" is better from some sites, usually on mountaintops, than it is on others, but it never goes away.

However, cosmic objects may be very large but very faint. For example, the Andromeda Galaxy, the nearest major galaxy to our own, has an apparent size in the sky seven times that of the full Moon, but is so far away and faint that only the brightest portion of its central "nucleus" can be seen with the naked eye, and then only on a dark night. That means light collection is actually more important than magnification.

The light collecting ability of a telescope is a function of the telescope's "aperture", or total amount of light collecting area, which for a refractor is the area of the objective lens. Magnification is simply a function of the eyepiece. Since the area of a disk is proportional to the square of the diameter of a disk, doubling the linear dimensions of an objective lens quadruples the area and light-gathering capability. A refractor with a 20 centimeter objective lens has four times the light-gathering capability of a similar instrument with a 10 centimeter lens.

From the beginning of telescopic astronomy, astronomers have sought bigger apertures. Bigger and bigger refractors were built, leading to the Yerkes refractor, which went online in 1897 and had an objective lens a bit over a meter in diameter. Nobody ever built a bigger refractor. Larger lenses were just too hard to grind, and it was difficult to fabricate thick, clear glass; worst of all, a huge lens tends to warp under its own weight. Modern large telescopes are "reflectors", using mirrors, discussed below. However, refractors still remain in widespread use in small telescopes and binoculars.

- One interesting variation on the refracting telescope is the door peephole, which allows a homeowner to see who's standing outside at the door. Such a peephole consists of a "fisheye" lens and a convex eyepiece lens. The fisheye lens has an exaggerated curvature on the outer side that bends light rays inward over a very wide field of view. Since that by itself would result in an impractically short focal length, the inner side is concave to bend the accumulated light rays outward again and focus them on the eyepiece lens so they can be viewed.
- One of the most common optical technologies, aside from eyeglasses, is the camera. Modern cameras in general consist of a lens system focusing an image on plastic film coated with photosensitive dyes, or in modern times, an electronic imaging sensor array that produces a "digital" image. Both films and imaging sensors are complicated subjects by themselves; they are generally ignored here to focus on camera optics.

The camera was invented in the 19th century. 19th century cameras originally used photographic plates instead of flexible film, with the plate made of glass covered with photosensitive dyes. A film plate could be placed inside a pinhole camera to obtain an image, but it was very difficult to obtain a *good* image with such crude technology. The first really practical cameras used a convex lens to focus light on a film plate, with an angled, partly transparent mirror in the optical path to reflect the image to a ground-glass plate that acted as a "viewfinder". The light from the lens was "gated" by a shutter system.

The photographer could hold the shutter open longer in faint light to get a good image, though the longer the shutter time, the harder it was to keep the image from being blurred due to motion during the "exposure". The lens of the camera was fitted on a "bellows" system that could be moved back and forth to adjust focus.

Late in the 19th century, the Kodak company introduced the simple portable "box camera", which had a fixed shutter speed and a fixed focus, allowing it to take adequate pictures of objects a few steps away in normal lighting conditions. The box camera, crude as it was, eliminated the monopoly of the professional cameraman on photography, allowing everyone to take pictures. It more or less survives today as the cheap "disposable camera", though some of these have more advanced features, and they use film on a plastic roll, a technology introduced in the 1920s that became all but standard until the intrusion of non-film digital electronic cameras in the 1990s.

A typical reasonable-quality modern film or digital electronic camera will feature:

- A "zoom lens", which is basically a refracting telescope that can be adjusted to give higher magnifications for an image.
- An "autofocus" system, which uses technologies such as a rangefinding light beam or an electronic system that adjusts the focus to obtain the sharpest edges in the image.
- Automatic light metering, with the camera increasing the shutter exposure time to adjust for low-light conditions.
- A "reflex" viewfinder, in which the angled partial mirror is pivoted out of the line of sight to the image when the shutter button is pressed.

Midrange or high-end cameras can use special lens assemblies, such as a heavily curved fisheye lens for wide-angle shots, that can be fitted over the regular camera lens assembly. Such modern "point and shoot" allow amateurs with modest experience to often take professional-quality images, though of course the professionals can obtain them much more regularly and under more difficult "shooting" conditions.

MIRRORS AND REFLECTING TELESCOPES

- The examples in this section so far have used lenses. Very similar principles apply to mirrors, though of course with mirrors the rays are reflected, not passed through.

The most familiar mirror is the flat bathroom mirror; rays bounce directly off of it. When Dexter looks into a bathroom mirror from a given distance away, the image of his face seems to be at an equal distance on the other side of the mirror.

This is handy for brushing his hair and the like. Flat mirrors can be arranged in a tube at oblique angles to see around corners. Such a device is known as a "periscope". Periscopes are of course familiar from their use in armored vehicles and particularly submarines. A classic submarine periscope features an evacuated tube and, since the tube tends to be long, telescopic elements. In modern times, submarine periscopes may actually carry electronic imagers, with the images relayed back to the submarine over wires.

Periscopic principles are also used in the "coude focus" for telescopes, which is a periscopic system that can rotate as the telescope tracks the sky to bounce the image onto a camera system, a spectroscope, or other instrument. In addition, simple telescopes use mirrored prisms to create "folded optics", reducing length; the best known example is the ordinary binocular, in which the mirrored prisms bounce the dual images obtained from the separated objective lenses to the closely-separated eyepieces.

- Mirrors can also have concave or convex surfaces. Common handheld "shaving mirrors" will have a flat mirror on one side and a concave mirror on the other; the concave surface will move the focus towards the surface of the mirror, with this "leveraging" providing magnification. When Dexter looks into the concave mirror, the image of his face appears to be closer on the far side of the mirror.

A SHAVING MIRROR (GVG / PD)

In contrast, the fisheye mirrors sometimes used on the side rear-view mirrors of trucks and the like are convex, extending the focus to allow a wider field of view at the expense of some reduction in the size of the image. If Dexter looks into a convex mirror, the image of his face appears to be farther away on the far side of the mirror. Fans of Gary Larson's FAR SIDE will remember one cartoon in which a woman driver sees a huge eye filling up her side rear-view mirror, which is marked with the text: Objects are nearer than they appear.

One of the advantages of mirrors is that all wavelengths of visible light reflect in the same way, and so chromatic aberration doesn't occur.

In Newton's day, the compound lens system used to cancel chromatic aberration hadn't been invented. Since Newton didn't realise that different materials have different indexes of refraction, he didn't think there was any way to get rid of chromatic aberration in a refracting telescope, and so he invented a "reflective" telescope or "reflector" with a concave mirror at the bottom to collect and focus light.

Newton's form of reflector became known, of course, as "Newtonian"; this form involves reflecting the light from the main "primary" mirror to a small "secondary" mirror mounted in the front end of the telescope, with the secondary mirror set at an angle to reflect the image to an eyepiece on the side. The main alternate form in modern times is "Cassegrain" telescope, in which a secondary mirror with a hyperbolic surface reflects the light back down through a hole in the primary mirror into an eyepiece.

Difficulties in making mirrors that could keep their form and remain shiny made reflectors inconvenient until late in the 19th century, when techniques were discovered for laying down a thin layer of silver on a glass mirror. In the next century, improved techniques were developed where aluminum was laid down on a mirror blank as a hot metal vapour in a vacuum chamber. Since it is very difficult to build very large lenses, reflectors are now the norm for large telescopes, with apertures far larger than any refractor ever built.

- One of the problems with a standard reflecting telescope is that its field of view is fairly narrow; the parabolic curvature of the mirror elongates the part of the image at the edges of the field of view, a phenomenon known as "coma". Wide-field telescopes generally use a "hybrid" reflector-refractor configuration known as a "Schmidt telescope".

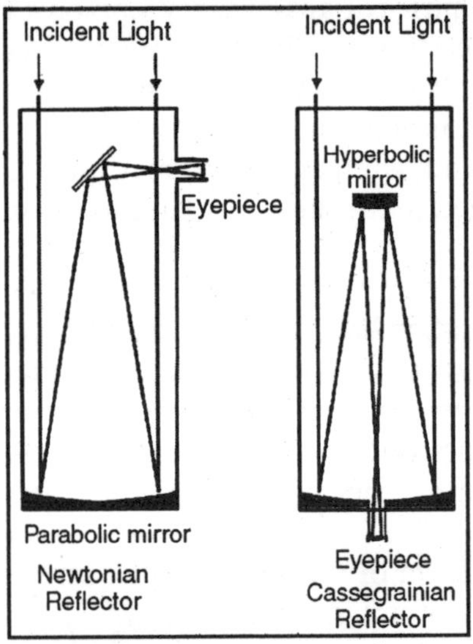

The Schmidt features a primary mirror with a spherical curvature, which would normally give spherical aberration, but the front of the telescope is fitted with a "corrector" lens thick on the edge, thick in the centre, thin in between to adjust for spherical aberration. The "Schmidt-Cassegrain" configuration, with a hole in the primary mirror and the eyepiece in back, is a popular design for compact amateur telescopes.

- Through most of the 20th century, most large astronomical telescopes recorded their images on photographic film. This not only allowed recording the images, permitting inspection at leisure and long-term archiving, but the photographic film could build up faint images over long exposures. Modern large astronomical telescopes now use electronic sensor systems, linked to computers that store and analyse the image data. Modern sensors and computer power have been able to compensate for the "seeing" problem to an extent, allowing large telescopes to obtain more detail.

Telescopes carried on orbiting astronomical satellites, such as the famous NASA Hubble Space Telescope, may have a "turntable" that carries different sensors, for example one with a wide field of view and coarse angular resolution, another with a narrow field of view and high angular resolution, and a third that performs spectroscopy. Other switching schemes may be used to divert a space telescope's image if the sensors are too large to be mounted on a turntable.

- It is difficult to grind the large mirrors for big astronomical telescopes, and traditionally they have had to be made out of glasses that have a low coefficient of thermal expansion, such as Pyrex and in particular a glass known as "Zerodur", since otherwise their precision surfaces would shift in shape with temperature, distorting their images. Large telescopes usually need a period to cool down at night before they can perform observations.

In fact, after the construction of the 5.1 meter Hale telescope at Mount Palomar observatory in the late 1940s, there was a general belief that telescopes couldn't get much bigger, because the thermal mass of the mirror would become so great that the mirror wouldn't be able to cool down and stabilize until the night was over. The mirror would also be impractically heavy to mount and use.

However, in the 1980s and 1990s new telescopes were designed that got around this limitation. New lightweight mirror materials, along with improved design and fabrication methods, mean that it is now practical to build mirrors up to eight meters in diameter. Even larger telescopes can now be made as assemblies of mirrors. The famous pair of Keck telescopes on Mauna Kea in Hawaii each use a set of 36 hexagonal thin mirror segments that are fitted together on a mechanical support structure. The support structure continuously adjusts the positions of the segments to ensure a good focus, with the adjustments controlled by optical sensors.

Each Keck telescope has a composite aperture of 10.1 meters, with four times the light-gathering power of the Hale telescope. This is by no means the end of the line, either; European astronomers are pushing for construction of a 42 meter reflecting telescope using mosaic mirror technology. They originally wanted a 100 meter telescope, but though it was judged feasible the expense was too great.

RADIO, INFRARED, ULTRAVIOLET, AND × RAY TELESCOPES

- Optical telescopes have led to telescopes to examine other regions of the spectrum. A typical radio telescope focuses long-wavelength radio signals using a large dish. The dish does not have to be as precisely built since the mirror of an optical telescope because the surface precision required increases as the wavelength of the electromagnetic radiation decreases, and radio wavelengths are very long compared to visible light wavelengths.

Infrared telescopes use more or less conventional optical telescope designs, but since infrared radiation is mostly absorbed by the atmosphere, infrared telescopes are usually built to be flown in balloons, high-flying aircraft, or as orbital satellites. They have to be cooled, traditionally with liquid helium in a double-dewar flask, to be sensitive to faint sources of radiation.

Short-wavelength high ultraviolet and×ray radiation are also mostly absorbed by the atmosphere and have to be observed by satellites. These short wavelengths cannot be focused by a conventional mirror, since they are so energetic that they go right into it. However, they will "skip" over a mirror at low angles, like a flat rock skipping over the surface of water, and so they can be focused by "grazing incidence" mirrors, which look like tubes that taper to a smaller diameter in a smooth curve from front to back. Such grazing-incidence mirrors are nested to increase the amount of radiation they focus. Typically, they consist of a primary stage with four nested cylinders with a parabolic curvature, feeding a secondary stage with four nested cylinders with hyperbolic curvature.

The secondary stage focuses the radiation on an electronic detector array, which has to be shielded with lead along its bottom and sides to prevent stray×rays from introducing noise. The problem with the grazing incidence scheme is that the skipping angle gets shallower and shallower as the wavelength of the×rays increases and the×rays become more energetic, resulting in a very long telescope.

Another scheme, known as a "coded mask" telescope, is used for short-wavelength×rays. A coded mask×ray telescope is essentially a pinhole camera, with a heavy metal plate perforated with holes mounted above a detector array. Multiple holes are used to increase the flux of×rays to the detector array; the holes are carefully arranged so that the actual×ray image can be sorted out with a bit of computer power.

A coded mask telescope can also focus low-energy gamma rays, but even it will not handle high energy gamma rays. They must be measured by "particle detector" systems that are way beyond the scope of this document.

- Since the 1960s, radio telescopes have been linked together to form radio interferometers that can compare the phase of radio waves from distant cosmic objects, providing very high resolutions. The resolution of a radio interferometer array is proportional to the spacing or "baselines" of the radio telescopes that make up the array. Radio telescopes have been linked up over continents to provide extreme resolution.

  Notice that the length of the baselines in the array only affects the resolution of the images obtained. The sensitivity of the array, or essentially the amount of radio energy it can collect, remains proportional to the sum of the areas of the individual radio telescopes in the array.

In the late 1980s, optical interferometer telescope arrays

have been built as well. Due to the fact that optical wavelengths are much shorter than radio wavelengths, an optical interferometer requires extremely precise spacing and separations of present systems are no more than a few hundred meters.

FOOTNOTE: STEREO IMAGING SCHEMES

- One of the interesting and amusing examples of optical technology is stereo imaging. Humans and most other primates, like monkeys and apes, have stereo vision; their eyes are set side-by-side in the face, with both eyes covering the same view but at a slightly different angle. The brain's visual processing system combines the two views to give information on the distance of objects. Animals, such as horses and many species of fish, that have eyes more to the side of the head have a wider field of view but only a limited stereo capability, directly ahead.

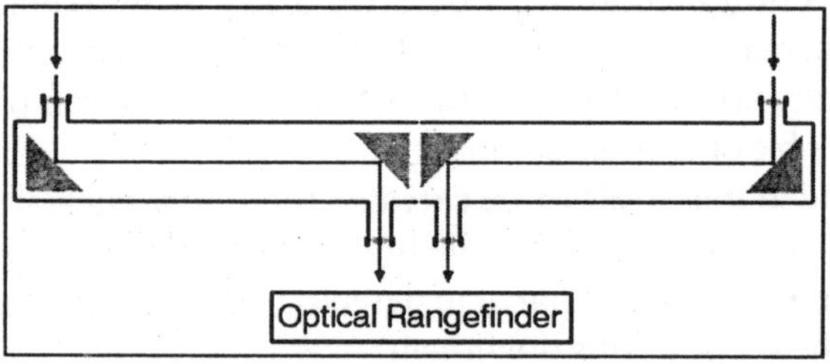

Old-style optical rangefinders, used on warships to find the range for naval guns, were based on exaggerating stereo vision. Such an optical rangefinder was essentially two telescopic periscopes with the objective lenses spaced widely apart, left to right. To estimate range, a sailor looked through the two eyepieces and adjusted a knob to bring the left and right views of the target together. Turning the knob also changed a range reading corresponding to the relative angle of the two objective lenses. Matching up the two images then gave the range to the target.

The basic idea behind stereo imaging is somewhat reversed from that of an optical rangefinder. The trick is to create two or more images of the same object, observed with an offset, and pass them to each eye in a superimposed fashion, with the brain's optical system merging them and producing an impression of depth. The crudest way to do this is to simply print the offset images side by side, and then provide registration marks so that readers can cross their eyes appropriately and get the proper stereo effect. This is a uncomfortable scheme, and it is much easier to use a viewer, with a lens for each eye that bends each image to each eye so the eyes don't have to be bent themselves. This is the approach used in the old-fashion stereo viewers used a century ago, and their modern "ViewMaster" equivalent. An acquaintance of mine makes his own stereo images by taking a picture of, say, a building, then stepping several paces to one side and taking another picture of the building. He then pastes the prints onto cardboard to display using an old-time stereo viewer.

This technology has serious, even deadly serious, applications as well. Earth surveillance satellites, both for civil and military surveillance purposes, sometimes carry stereo camera gear, with one camera taking a picture of a target before the satellite passes over it and a second camera taking a picture of the target after the satellite passes over it. The stereo images can be viewed through a stereo viewer to reveal details that might have gone unnoticed in a two-dimensional image.

The "lenticular" plastic stereo cards occasionally used as trinkets or promotional gimmicks don't require either crossing the eyes, or a viewer. They consist of a set of images cut into narrow slices and interleaved into a card, then overlaid with a plastic panel consisting of an array of lenses organized as long narrow vertical strips and with a half-circular cross-section. The lenses optically sort out one of the interleaved images, depending on viewing angle, with an offset image sent to each eye to produce the illusion of depth; rotate the card onto its side and the stereo effect goes away.

Of course, obtaining a good stereo image requires high quality printing and accurate registration with the lens sheet. Cheaper lenticular trinkets, which use rows of coarse plastic prisms, simply give a different image when shifted back and forth; for example, a kitschy old standard is to have a pretty girl who is fully dressed from one angle, but is wearing much less from another. Flat-panel computer displays are now available that use lenticular technology to provide a stereo image, with the computer software slicing up the image and interleaving it for display.

Another traditional stereo scheme is the "anaglyph", which consists of one offset image in red overprinted with the other in blue. Blue and red glasses are used to merge the images and get the stereo effect. This scheme was also used in the notoriously trashy 3D movies of the 1950s, but it had a major limitation: it could only be used with black-and-white videos. Modern colour stereo movies, such as those popular at theme parks, project the two images using two polarizations at 90 degrees to each other, and the glasses used by the audience similarly have two polarized lenses with the plane of polarization also at right angles to each other.

LIGHT AND A NEW PHYSICS

- The astute reader will have some questions about the glib statements made in this chapter and the previous chapter. For example, sound is a wave phenomenon, a transfer of mechanical energy through air or some other medium. But what is the medium through which light is transferred? It transfers through a vacuum that by definition contains absolutely nothing.

This was a puzzle to physicists before the 20th century, and it was one of the reasons that there was uncertainty as to whether light was a wave or particle until Thomas Young's two-slit experiment seemingly settled matters in favour of waves. A wave implies a transmission medium, while a particle flies across empty space just fine.

Since nobody could actually detect any medium that could propagate light waves, physicists had to imagine an invisible "luminiferous (light-bearing) ether" that permeated the Universe to do this job, but this was such an arbitrary construct that many were uncomfortable with it. It also led to another troublesome issue: if there was such an "ether" filling the Universe, how would the motion of the Earth, or any other object, affect the perceived velocity of light?

The great German-American physicist Albert Einstein went a bit further and wondered if were possible that the Earth or some other object might actually travel through the ether so fast that it would exceed the speed of light. Under such circumstances, a person would not be able to see his or her reflection in a mirror. Einstein moved from this line of reasoning to create his "Theory of Special Relativity", the first of the two major revolutions in physics in the 20th century.

This theory postulated that there was no ether, and that the speed of light was always the same to all observers, no matter what their speeds relative to each other might be. This led to some very non-intuitive consequences.

- The fact that there was no ether also tended to undermine the wave theory of light. This was baffling, since light clearly seemed to exhibit wave properties, such as interference. There were other difficulties as well. For example, if a thermal emitter produced light of wavelengths all up the EM spectrum, the energy would have been infinite, destroying the Universe in the flare of a single match.

Instead, a thermal emitter produces a black body spectrum, with the energy falling off above the peak emission wavelengths. Another issue was the nice neat spectral lines of light emitted or absorbed by atoms. It seemed like atoms were some sort of resonant chambers for light, but nobody could figure out exactly why that would be so.

Yet another puzzle was light polarization. As mentioned, if we put two polarized lenses in series and at right angles to each other, all the light is blocked out; but if we put *another* polarized lens between the two at a 45-degree angle, light will be attenuated but still filter through. There was no way to explain this in classical physics.

These and other puzzles led to the second major revolution in 20th century physics, known as "quantum physics". Quantum physics led to even more bizarre quandaries than relativistic physics, proclaiming that particles could be waves, waves could be particles, and that there were fundamental limits on how well physical interactions could be measured.

Both relativistic and quantum physics broke the boundaries of conventional physics, and as such they leave the domain of this document. Relativistic and quantum physics are discussed in companion documents to this one.

Chapter 13

Special Relativity

SYMMETRY OF RELATIVITY

Maxwell's equations are invariant under Lorentz transformations, as well as translations and spatial rotations. Together these transformations comprise the Poincare group. Of course, Maxwell's equations are also invariant under spatial and temporal reflections, but it is often overlooked that in addition to all these linear transformations, Maxwell's equations possess still another symmetry, namely, the symmetry of spacetime *inversion.*

In a sense, an inversion is a kind of reflection about a surface in spacetime, analogous to inversions about circles in projective geometry, the only difference being that the Minkowski interval is used instead of the Euclidean line element.

Consider two events E_1 and E_2 that are null-separated from each other, meaning that the absolute Minkowski interval between them is zero in terms of an inertial coordinate system x, y, z, t. Let s_1 and s_2 denote the absolute intervals from the origin to these two events (respectively).

Under an inversion of the coordinate system about the surface at an absolute interval R from the origin (which may be chosen arbitrarily), each event located on a given ray through the origin is moved to another point on that ray such that its absolute interval from the origin is changed from s to R^2/s. Thus the hyperbolic surfaces outside of R are mapped to surfaces inside R, and vice versa. To prove that two events originally separated by a null Minkowski interval are still null-separated after the coordinates have been inverted, note that the ray from the origin to the event E_j can be characterized by constants α_j, β_j, γ_j defined by

$$x_j - \alpha_j t_j \qquad y_j - \beta_j t_j \qquad z_j - \gamma_j t_j$$

In terms of these parameters the magnitude of the interval from the origin to E_j can be written as

$$s_j = t_j\sqrt{\alpha_2^2 + \beta_2^2 + \gamma_2^2 - 1}$$

The squared interval between E_1 and E_2 can then be expressed as

$$\begin{aligned} s_{12}^2 &= (x_2 - x_1)^2 + (y_2 - y_1)^2 + (z_2 - z_1)^2 - (t_2 - t_1)^2 \\ &= \left(\alpha_2^2 + \beta_2^2 + \gamma_2^2 - 1\right)t_2^2 + \left(\alpha_1^2 + \beta_1^2 + \gamma_1^2 - 1\right)t_1^2 \\ &\qquad -2\left(\alpha_1\alpha_2 + \beta_1\beta_2 + \gamma_1\gamma_2 - 1\right)t_1 t_2 \\ &= s_1^2 + s_1^2 - 2K_{12}s_1 s_2 \end{aligned}$$

where

$$K_{12} = \frac{\left(\alpha_1\alpha_2 + \beta_1\beta_2 + \gamma_1\gamma_2 - 1\right)}{\sqrt{\left(\alpha_1^2 + \beta_1^2 + \gamma_1^2 - 1\right)\left(\alpha_2^2 + \beta_2^2 + \gamma_2^2 - 1\right)}}$$

Since inversion leaves each event on its respective ray, the value of K_{12} for the inverted coordinates is the same as for the original coordinates, so the only effect on the Minkowski interval between E_1 and E_2 is to replace s_1 and s_2 with R^2/s_1 and R^2/s_2 respectively. Therefore, the squared Minkowski interval between the two events in terms of the inverted coordinates is

$$\left(s_{12}{}^t\right)^2 = \frac{R^2}{s_1^2} + \frac{R^2}{s_2^2} - 2K_{12}\frac{R^2}{s_1 s_2} = \frac{R^2}{s_1 s_2}\left(s_1^2 + s_2^2 - 2K_{12}s_1 s_2\right)$$

The quantity in parentheses on the right side is just the original squared interval, so if the interval was zero in terms of the original coordinates, it is zero in terms of the inverted coordinates. Thus inversion of a system of inertial coordinates yields a system of coordinates in which all the null intervals are preserved.

It was shown in 1910 by Bateman and (independently) Cunningham that this is the necessary and sufficient condition for Maxwell's equations to be invariant. Incidentally, Einstein was dismissive of this invariance when Felix Klein asked him about it. He wrote I am convinced that the covariance of Maxwell's formulas under transformation according to reciprocal radii can have no deeper significance; although this transformation retains the form of the equations, it does not uphold the correlation between coordinates and the measurement results from measuring rods and clocks.

Einstein was similarly dismissive of Minkowski's "formal approach" to spacetime at first, but later came to appreciate the profound significance of it. In any case, it's interesting to note that straight lines in inertial coordinate systems map to straight or hyperbolic paths under inversion. This partly accounts for the fact that, according to the Lorentz-Dirac equations of classical electrodynamics, perfect hyperbolic motion *is* inertial motion, in the sense that there are free-body solutions describing particles in hyperbolic motion, and a charged particle in hyperbolic motion does not radiate.

It's also interesting that the relativistic formula for composition of two speeds is invariant under inversion of the arguments about the speed c, i.e., replacing each speed v with c^2/v.

Letting f(u, v) denote the composition of the (co-linear) speeds u and v, and choosing units so that c = 1, we can impose the three requirements

$$0 = f(u,-u) \qquad u = f(0,u) = f(u,0) \qquad f(u,v) = f(1/u,1/v)$$

The first two requirements are satisfied by both the Galilean and the Lorentzian composition formulas, but the third requirement is not satisfied by the Galilean formula, because that gives

$$f(u,v) = u + f \qquad f(1/u,1/v) = \frac{1}{u} + \frac{1}{v} = \frac{u+v}{uv}$$

However, somewhat surprisingly, the relativistic composition function gives

$$f(u,v) = \frac{u+v}{1+uv} \qquad f(1/u,1/v) = \frac{\frac{1}{u}+\frac{1}{v}}{\frac{1}{u}\frac{1}{v}+1} = \frac{v+u}{1+uv}$$

so it does comply with all three requirements. This singles out the composition law with k = 1 from the previous chapter. As indicated by Einstein's reply to Klein, the physical significance of such inversion symmetries is obscure, and we should also note that the spacetime inversion is not equivalent to the speed inversion, although they are formally very similar. To clarify how this symmetry arises in the relativistic context, recall that we had derived at the end of the previous chapter the relation

$$k\,uvw + u + v + w = 0$$

where $u = v_{12}$, $v = v_{23}$, and $w = v_{31}$. The symbol v_{ij} signifies the speed of the ith particle in terms of the inertial rest frame coordinates of the jth particle. With k = 0 this corresponds to the Galilean speed composition formula, which clearly is not invariant under inversion of any or all of the speeds. For any non-zero value of k, equation can be re-written in the form

$$\left(1+u\sqrt{k}\right)\left(1+v\sqrt{k}\right)\left(1+w\sqrt{k}\right)=\left(1-u\sqrt{k}\right)\left(1-v\sqrt{k}\right)\left(1-w\sqrt{k}\right)$$

Squaring both sides of this equation gives the equality

$$\left(1+u\sqrt{k}\right)^2\left(1+v\sqrt{k}\right)^2\left(1+w\sqrt{k}\right)^2 = \left(1-u\sqrt{k}\right)^2\left(1-v\sqrt{k}\right)^2\left(1-w\sqrt{k}\right)^2$$

If we replace each speed with its inversion in this formula, and then multiply through by $(uvw)^2 / k^3$ we get

$$\left(1+\frac{u}{\sqrt{k}}\right)^2\left(1+\frac{v}{\sqrt{k}}\right)^2\left(1+\frac{w}{\sqrt{k}}\right)^2=\left(1-\frac{u}{\sqrt{k}}\right)^2\left(1-\frac{v}{\sqrt{k}}\right)^2\left(1-\frac{w}{\sqrt{k}}\right)^2$$

which is equivalent to the preceding formula if and only if

$$\sqrt{k} = \pm\frac{1}{\sqrt{k}}$$

Hence the speed composition formula is invariant under inversion if k = ±1. The case k = −1 is equivalent to the case k = +1 if each speed is taken to be imaginary (corresponding to the use of an imaginary time axis), so without loss of generality we can choose k = +1 with real speeds. There remains, however, the ambiguity introduced by squaring both sides of equation, suppressing the signs of the factors.

Equation itself, without squaring, is invariant under inversion of any two of the speeds, but the inversion of all three speeds changes the sign of the right side. Thus by squaring both sides we make it consistent with either of the two complementary relations

$$uvw + u + v + w = 0 \qquad 1 + uv + uw + vw = 0$$

The left hand relation is invariant under inversion of any two of the speeds, whereas the right hand relation is invariant under inversion of one or all three of the speeds. The question, then, is why the first formula applies rather than the second. To answer this, we should first point out that, despite the formal symmetry of the quantities u, v, w in these equations, they are not conceptually symmetrical. Two of the quantities are implicitly defined in terms of one inertial coordinate system, and the third quantity is defined in terms of a different inertial coordinate system. In general, there are nine conceptually distinct speeds for three co-linear particles in terms of the three rest frame coordinate systems, namely

$$\begin{matrix} v_{11} & v_{12} & v_{13} \\ v_{21} & v_{22} & v_{23} \\ v_{31} & v_{32} & v_{33} \end{matrix}$$

where v_{ij} is the speed of the ith particle in terms of the inertial rest frame coordinates of the jth particle. By definition we have $v_{ii} = 0$ and by reciprocity we have $v_{ij} = -v_{ji}$, so the speeds comprise an anti-symmetric array.

Thus, although the three speeds v_{12}, v_{23}, v_{31} are nominally defined in terms of three different systems of coordinates, any two of them can be expressed in terms of a single coordinate system by invoking the reciprocity relation.

For example, the three quantities v_{12}, v_{23}, v_{31} can be expressed in the form v_{12}, $-v_{32}$, v_{31}, which signifies that the first two speeds are both defined in terms of the rest frame coordinates of frame 2. However, the remaining speed does not have a direct expression in terms of that frame, so a composition formula is needed to relate all three quantities. We've seen that the relativistic composition formula yields the same value for the third speed regardless of whether we use the two other speeds or their reciprocals. To more clearly exhibit the peculiar 2+1 symmetry of this velocity composition law, note that it can be expressed in multiplicative form as

$$\left(\frac{1+v_{12}}{1-v_{12}}\right)\left(\frac{1+v_{23}}{1-v_{23}}\right)\left(\frac{1+v_{31}}{1-v_{31}}\right)=1$$

where v_{ij} denotes the speed of object j with respect to object i. Clearly if we replace any two of the speeds with their reciprocals, the relation remains unchanged. On the other hand, if we replace just one or all three of the speeds with their reciprocals, their product is still unity, but the sign is negated. Thus, one way of expressing the full symmetry of this relation would be to square both sides, giving the result

$$\left(\frac{1+v_{12}}{1-v_{12}}\right)^2\left(\frac{1+v_{23}}{1-v_{23}}\right)^2\left(\frac{1+v_{31}}{1-v_{31}}\right)^2=1$$

which is completely invariant under any replacement of one or more speeds with their respective reciprocals. Naturally we can extend the product of factors of the form $(1+v_{ij})/(1-v_{ij})$ to any cyclical sequence of relative speeds between any number of co-linear points.

It's interesting to note the progression of relations between the speeds involving one, two, and three particles. The relativity of position is expressed by the identity

$$v_{ii}=0$$

for any one particle, and the relativity of velocity can be expressed by the skew symmetry

$$v_{ij}+v_{ji}=0$$

for any two systems particles. (This was referred to earlier as the reciprocity condition $v_{ij}=-v_{ji}$). The next step is to consider the cyclic sum involving three particles and their respective inertial rest frame coordinate systems. This is the key relation, because all higher-order relations can be reduced to this. If acceleration were relative (like position and velocity), we would expect the cyclic symmetry $v_{ij}+v_{jk}+v_{ki}=0$, which is a linear function of all three components. Indeed, this is the Galilean composition formula. However, since acceleration is absolute, it's to be expected that the actual relation is non-linear in each of the three components. So, instead of vanishing, we need the right side of this sum to be a symmetric function of the terms. The only other odd elementary symmetric function of three quantities is the product of all three, so we're led (again) to the relation

$$v_{ij}+v_{jk}+v_{ki}=-v_{ij}v_{jk}v_{ki}$$

which can be regarded as the law of inertia. Since there is only one odd elementary symmetric function of one variable, and likewise for two variables, the case of three variables is the first for which there exists a non-tautological expression of this form.

We may also note a formal correspondence with De Morgan's law for logical statements. Letting sums denote logical ORs (unions), products denote logical ANDs (intersections), and overbars denote logical negation, De Morgan's law states that

$$\overline{X+Y+Z} = \overline{X}\ \overline{Y}\ \overline{Z}$$

for any three logical variables X, Y, Z. Now, using the skew symmetry property, we can "negate" each velocity on the right hand side of the previous expression to give

$$v_{ij} + v_{jk} + v_{ki} = v_{ji} v_{kj} v_{ik}$$

From this standpoint the right hand side is analogous to the "logical negation" of the left hand side, which makes the relation analogous to setting the quantity equal to zero.

The justification for regarding this relation as the source of inertia becomes, which describes how the relativistic composition law for velocities accounts for the increasing inertia of an accelerating object. This leads to the view that inertia itself is, in some sense, a consequence of the non-linearity of velocity compositions. Given the composition law u' = (u+v)/(1+uv) for co-linear speeds, what can we say about the transformation of the coordinates x and t themselves under the action of the velocity v? The composition law can be written in the form vuu'+u'–u = v, which has a natural factorization if we multiply through by v and subtract 1 from both sides, giving

$$(1 + vu)(1 - uv') = 1 - v^2$$

If u and u' are taken to be the spatio-temporal ratios x/t and x'/t', the above relation can be written in the form

$$\left(1 + v\frac{x}{t}\right)\left(1 - v\frac{x'}{t'}\right) = 1 - v^2$$

On the other hand, remembering that we can insert the reciprocals of any two of the quantities u, u', v without disturbing the equality, we can take u and u' to be the temporal-spatial ratios t/x and t'/x' to give

$$\left(1 + v\frac{t}{x}\right)\left(1 - v\frac{t'}{x'}\right) = 1 - v^2$$

These last two equations immediately give

$$t't = \frac{(t + vx)(t' + vx')}{1 - v^2} \qquad x'x = \frac{(x + vt)(x' + vt')}{1 - v^2}$$

Treating the primed and unprimed frames equivalently, and recalling that v' = –v, we see that a perfectly symmetrical factorization, so we exploit this factorization to give the transformation equations

$$t' = \frac{t + vx}{\sqrt{1 - v^2}} \qquad x' = \frac{x + vt}{\sqrt{1 - v^2}}$$

$$t = \frac{t' + vx'}{\sqrt{1 - v^2}} \qquad x = \frac{x' + vt'}{\sqrt{1 - v^2}}$$

These are the Lorentz transformations for velocity v in the x direction. The y and z coordinates are unaffected, so we have y' = y and z' = z.

From this it follows that the quantity $t^2 - x^2 - y^2 - z^2$ is invariant under a general Lorentz transformation, so we have arrived at the full Minkowski spacetime metric.

Now, to determine the full velocity composition law for two systems of aligned coordinates k and K, the latter moving in the positive x direction with velocity v relative to the former, we can without loss of generality make the origins of the two systems both coincide with a point P_0 on the subject worldline, and let P_1 denote a subsequent point on that worldline with k system coordinates dt, dx, dy, dz.

By definition the velocity components of that worldline with respect to k are $u_x = dx/dt$, $u_y = dy/dt$, and $u_z = dz/dt$. The coordinates of P_1 with respect to the K system are given by the Lorentz transformation for a simple boost v in the x direction:

$$dX = \frac{dx - vdt}{\gamma} \quad dY = dy \quad dZ = dz \quad dT = \frac{dt - vdx}{\gamma}$$

where $\gamma = \sqrt{1 - v^2}$. Therefore, the velocity components of the worldline with respect to the K system are

$$U_X = \frac{dx - vdt}{dt - vdx} = \frac{u_x - v}{1 - vu_x}$$

$$U_Y = \frac{\gamma dy}{(dt - vdx)} = \frac{\gamma u_y}{(1 - vu_x)}$$

$$U_Z = \frac{\gamma dz}{(dt - vdx)} = \frac{\gamma u_z}{(1 - vu_x)}$$

FRAMES OF REFERENCE AND NEWTON'S LAWS

The cornerstone of the theory of special relativity is the Principle of Relativity: The Laws of Physics are the same in all inertial frames of reference. We shall see that many surprising consequences follow from this innocuous looking statement. Let us first, however, briefly review Newton's mechanics in terms of frames of reference.

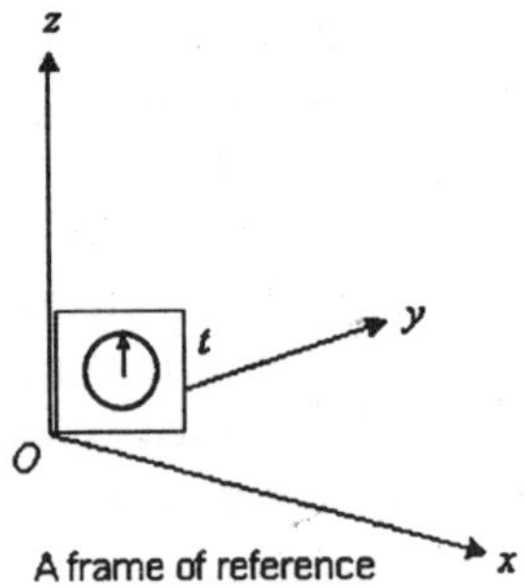

A frame of reference

A "frame of reference" is just a set of coordinates: something you use to measure the things that matter in Newtonian problems, that is to say, positions and velocities, so we also need a clock. A point in space is specified by its three coordinates (x, y, z) and an "event" like, say, a little explosion, by a place and time: (x, y, z, t). An inertial frame is defined as one in which Newton's law of inertia holds—that is, any body which isn't being acted on by an outside force stays at rest if it is initially at rest, or continues to move at a constant velocity if that's what it was doing to begin with.

An example of a *non*-inertial frame is a rotating frame, such as a carousel. The "laws of physics" we shall consider first are those of Newtonian mechanics, as expressed by Newton's Laws of Motion, with gravitational forces and also contact forces from objects pushing against each other.

For example, knowing the universal gravitational constant from experiment (and the masses involved), it is possible from Newton's Second Law,

$$force = mass \times acceleration,$$

to predict future planetary motions with great accuracy.

Suppose we know from experiment that these laws of mechanics are true in one frame of reference. How do they look in another frame, moving with respect to the first frame? To find out, we have to figure out how to get from position, velocity and acceleration in one frame to the corresponding quantities in the second frame.

Obviously, the two frames must have a constant relative velocity, otherwise the law of inertia won't hold in both of them. Let's choose the coordinates so that this velocity is along the x-axis of both of them.

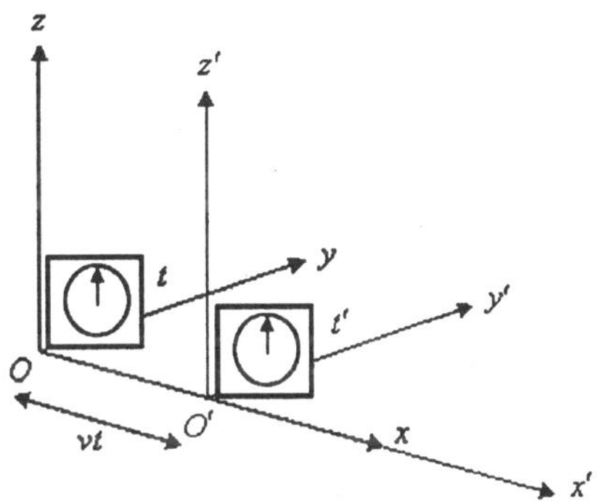

Fig. Two Frames of Reference Relatively Dispalced Along the x-axis

Notice we also throw in a clock with each frame.

Suppose S' is proceeding relative to S at speed v along the x-axis. For convenience, let us label the moment when O' passes O as the zero point of timekeeping.

Now what are the coordinates of the event (x, y, z, t) in S'? It's easy to see $t' = t$—we synchronized the clocks when O' passed O. Also, evidently, $y' = y$ and $z' = z$, from the figure. We can also see that $x = x' + vt$. Thus (x, y, z, t) in S corresponds to (x', y', z', t') in S', where

$$x' = x - vt$$

$$y' = y$$
$$z' = z$$
$$t' = t$$

That's how positions transform; these are known as the Galilean transformations. What about velocities? The velocity in S′ in the x′ direction

$$u_x' = \frac{dx'}{dt'} = \frac{dx'}{dt} = \frac{d}{dt}(x - vt) = \frac{dx}{dt} - v = u_x - v.$$

This is obvious anyway: it's just the addition of velocities formula

$$u_x = u_x' + n.$$

How does acceleration transform?

$$\frac{du_x'}{dt'} = \frac{du_x'}{dt} = \frac{d}{dt}(u_x - v) = \frac{du_x}{dt}$$

since *v* is constant. That is to say,

$$a_x' = a_x$$

the *acceleration is the same* in both frames. This again is obvious—the acceleration is the rate of change of velocity, and the velocities of the same particle measured in the two frames differ by a *constant* factor-the relative velocity of the two frames. If we now look at the motion under gravitational forces, for example,

$$m_1\vec{a} = \frac{Gm_1m_2}{r^2}\hat{r}$$

we get the same law on going to another inertial frame because *every term in the above equation stays the same.*

Note that $m\vec{a}$ is the rate of change of momentum—this is the same in both frames. So, in a collision, say, if total momentum is conserved in one frame (the sum of individual rates of change of momentum is zero) the same is true in *all* inertial frames.

THE SPEED OF LIGHT

Attempts to measure the speed of light played an important part in the development of the theory of special relativity, and, indeed, the speed of light is central to the theory. The first recorded discussion of the speed of light is in Aristotle, where he quotes Empedocles as saying the light from the sun must take some time to reach the earth, but Aristotle himself apparently disagrees, and even Descartes thought that light traveled instantaneously. Galileo, unfairly as usual, in *Two New Sciences* has Simplicio stating the Aristotelian position,

Everyday experience shows that the propagation of light is instantaneous; for when we see a piece of artillery fired at great distance, the flash reaches our eyes without lapse of time; but the sound reaches the ear only after a noticeable interval.

Of course, Galileo points out that in fact nothing about the speed of light can be deduced from this observation, except that light moves faster than sound. He then goes on to suggest a possible way to measure the speed of light. The idea is to have two people far away from each other, with covered lanterns.

One uncovers his lantern, then the other immediately uncovers his on seeing the light from the first. This routine is to be practiced with the two close together, so they will get used to the reaction times involved, then they are to do it two or three miles apart, or even further using telescopes, to see if the time interval is perceptibly lengthened. Galileo claims he actually tried the experiment at distances less than a mile, and couldn't detect a time lag. From this one can certainly deduce that light travels at least ten times faster than sound.

MEASURING THE SPEED OF LIGHT WITH JUPITER'S MOONS

The first real measurement of the speed of light came about half a century later, in 1676, by a Danish astronomer, Ole Rφmer, working at the Paris Observatory. He had made a systematic study of Io, one of the moons of Jupiter, which was eclipsed by Jupiter at regular intervals, as Io went around Jupiter in a circular orbit at a steady rate. Actually, Rφmer found, for several months the eclipses lagged more and more behind the expected time, until they were running about eight minutes late, then they began to pick up again, and in fact after about six months were running eight minutes early.

The cycle then repeated itself. Rφmer realized the significance of the time involved-just over one year. This time period had nothing to do with Io, but was the time between successive closest approaches of earth in its orbit to Jupiter. The eclipses were furthest behind the predicted times when the earth was furthest from Jupiter.

The natural explanation was that the light from Io (actually reflected sunlight, of course) took time to reach the earth, and took the longest time when the earth was furthest away. From his observations, Rφmer concluded that light took about twenty-two minutes to cross the earth's orbit. "For it is now certain from the phenomena of Jupiter's satellites, confirmed by the observations of different astronomers, that light is propagated in succession and requires about seven or eight minutes to travel from the sun to the earth." This is essentially the correct value.

Of course, to find the speed of light it was also necessary to know the distance from the earth to the sun. During the 1670's, attempts were made to measure the parallax of Mars, that is, how far it shifted against the background of distant stars when viewed simultaneously from two different places on earth at the same time.

This (very slight) shift could be used to find the distance of Mars from earth, and hence the distance to the sun, since all *relative* distances in the solar system had been established by observation and geometrical analysis. According to Crowe, they concluded that the distance to the sun was between 40 and 90 million miles.

Measurements presumably converged on the correct value of about 93 million miles soon after that, because it appears Romer (or perhaps Huygens, using Rφmer's data a short time later) used the correct value for the distance, since the speed of light was calculated to be 125, 000 miles per second, about three-quarters of the correct value of 186, 300 miles per second. This error is fully accounted for by taking the time light needs to cross the earth's orbit to be twenty-two minutes instead of the correct value of sixteen minutes.

Starlight and Rain

The next substantial improvement in measuring the speed of

light took place in 1728, in England. An astronomer James Bradley, sailing on the Thames with some friends, noticed that the little pennant on top of the mast changed position each time the boat put about, even though the wind was steady. He thought of the boat as the earth in orbit, the wind as starlight coming from some distant star, and reasoned that the apparent direction the starlight was "blowing" in would depend on the way the earth was moving. Another possible analogy is to imagine the starlight as a steady downpour of rain on a windless day, and to think of yourself as walking around a circular path at a steady pace.

The apparent direction of the incoming rain will not be vertically downwards-more will hit your front than your back. In fact, if the rain is falling at, say, 15 mph, and you are walking at 3 mph, to you as observer the rain will be coming down at a slant so that it has a vertical speed of 15 mph, and a horizontal speed towards you of 3 mph.

Whether it is slanting down from the north or east or whatever at any given time depends on where you are on the circular path at that moment. Bradley reasoned that the apparent direction of incoming starlight must vary in just this way, but the angular change would be a lot less dramatic. The earth's speed in orbit is about 18 miles per second, he knew from Rφmer's work that light went at about 10, 000 times that speed.

That meant that the angular variation in apparent incoming direction of starlight was about the magnitude of the small angle in a right-angled triangle with one side 10, 000 times longer than the other, about one two-hundredth of a degree. Notice this would have been just at the limits of Tycho's measurements, but the advent of the telescope, and general improvements in engineering, meant this small angle was quite accurately measurable by Bradley's time, and he found the velocity of light to be 185, 000 miles per second, with an accuracy of about one per cent.

Fast Flickering Lanterns

The problem is, all these astronomical techniques do not have the appeal of Galileo's idea of two guys with lanterns. It would be reassuring to measure the speed of a beam of light between two points on the ground, rather than making somewhat indirect deductions based on apparent slight variations in the positions of stars.

We can see, though, that if the two lanterns are ten miles apart, the time lag is of order one-ten thousandth of a second, and it is difficult to see how to arrange that. This technical problem was solved in France about 1850 by two rivals, Fizeau and Foucault, using slightly different techniques. In Fizeau's apparatus, a beam of light shone between the teeth of a rapidly rotating toothed wheel, so the "lantern" was constantly being covered and uncovered.

Instead of a second lantern far away, Fizeau simply had a mirror, reflecting the beam back, where it passed a second time between the teeth of the wheel. The idea was, the blip of light that went out through one gap between teeth would only make it back through the same gap if the teeth had not had time to move over significantly during the round trip time to the far away mirror. It was not difficult to make a wheel with a hundred teeth, and to rotate it hundreds of times a second, so the time for a tooth to move over could be arranged to be a fraction of one ten thousandth of a second.

The method worked. Foucault's method was based on the same general idea, but instead of a toothed wheel, he shone the beam on to a rotating mirror. At one point in the mirror's rotation, the reflected beam fell on a distant mirror, which reflected it right back to the rotating mirror, which meanwhile had turned through a small angle. After this second reflection from the rotating mirror, the position of the beam was carefully measured.

This made it possible to figure out how far the mirror had turned during the time it took the light to make the round trip to the distant mirror, and since the rate of rotation of the mirror was known, the speed of light could be figured out. These techniques gave the speed of light with an accuracy of about 1, 000 miles per second.

Albert Abraham Michelson

Albert Michelson was born in 1852 in Strzelno, Poland. His father Samuel was a Jewish merchant, not a very safe thing to be at the time. Purges of Jews were frequent in the neighboring towns and villages. They decided to leave town. Albert's fourth birthday was celebrated in Murphy's Camp, Calaveras County, about fifty miles south east of Sacramento, a place where five million dollars worth of gold dust was taken from one four acre lot. Samuel prospered selling supplies to the miners. When the gold ran out, the Michelsons moved to Virginia City, Nevada, on the Comstock lode, a silver mining town. Albert went to high school in San Francisco. In 1869, his father spotted an announcement in the local paper that Congressman Fitch would be appointing a candidate to the Naval Academy in Annapolis, and inviting applications. Albert applied but did not get the appointment, which went instead to the son of a civil war veteran.

However, Albert knew that President Grant would also be appointing ten candidates himself, so he went east on the just opened continental railroad to try his luck. Unknown to Michelson, Congressman Fitch wrote directly to Grant on his behalf, saying this would really help get the Nevada Jews into the Republican party.

This argument proved persuasive. In fact, by the time Michelson met with Grant, all ten scholarships had been awarded, but the President somehow came up with another one. Of the incoming class of ninety-two, four years later twenty-nine graduated. Michelson placed first in optics, but twenty-fifth in seamanship.

The Superintendent of the Academy, Rear Admiral Worden, who had commanded the Monitor in its victory over the Merrimac, told Michelson: "If in the future you'd give less attention to those scientific things and more to your naval gunnery, there might come a time when you would know enough to be of some service to your country."

Michelson Measures the Speed of Light

On returning to Annapolis from the cruise, Michelson was commissioned Ensign, and in 1875 became an instructor in physics and chemistry at the Naval Academy, under Lieutenant Commander William Sampson. Michelson met Mrs. Sampson's niece,

Margaret Heminway, daughter of a very successful Wall

Street tycoon, who had built himself a granite castle in New Rochelle, NY. Michelson married Margaret in an Episcopal service in New Rochelle in 1877.

At work, lecture demonstrations had just been introduced at Annapolis. Sampson suggested that it would be a good demonstration to measure the speed of light by Foucault's method. Michelson soon realized, on putting together the apparatus, that he could redesign it for much greater accuracy, but that would need money well beyond that available in the teaching demonstration budget.

He went and talked with his father in law, who agreed to put up $2, 000. Instead of Foucault's 60 feet to the far mirror, Michelson had about 2, 000 feet along the bank of the Severn, a distance he measured to one tenth of an inch. He invested in very high quality lenses and mirrors to focus and reflect the beam. His final result was 186, 355 miles per second, with possible error of 30 miles per second or so. This was twenty times more accurate than Foucault, made the New York Times, and Michelson was famous while still in his twenties. In fact, this was accepted as the most accurate measurement of the speed of light for the next forty years, at which point Michelson measured it again.

THE MICHELSON-MORLEY EXPERIMENT

The Nature of Light

As a result of Michelson's efforts in 1879, the speed of light was known to be 186, 350 miles per second with a likely error of around 30 miles per second. This measurement, made by timing a flash of light travelling between mirrors in Annapolis, agreed well with less direct measurements based on astronomical observations. Still, this did not really clarify the *nature* of light.

Two hundred years earlier, Newton had suggested that light consists of tiny *particles* generated in a hot object, which spray out at very high speed, bounce off other objects, and are detected by our eyes. Newton's arch-enemy Robert Hooke, on the other hand, thought that light must be a kind of *wave motion*, like sound. To appreciate his point of view, let us briefly review the nature of sound.

The Wavelike Nature of Sound

Actually, *sound* was already quite well understood by the ancient Greeks. The essential point they had realized is that sound is generated by a vibrating material object, such as a bell, a string or a drumhead. Their explanation was that the vibrating drumhead, for example, alternately pushes and pulls on the air directly above it, sending out waves of compression and decompression (known as rarefaction), like the expanding circles of ripples from a disturbance on the surface of a pond.

On reaching the ear, these waves push and pull on the eardrum with the same frequency (that is to say, the same number of pushes per second) as the original source was vibrating at, and nerves transmit from the ear to the brain both the intensity (loudness) and frequency (pitch) of the sound.

There are a couple of special properties of sound waves (actually any waves) worth mentioning at this point. The first is called interference. This is most simply demonstrated with water waves. If you put two fingers in a tub of water, just touching the surface a foot or so apart, and vibrate them at the same rate to get two expanding circles of ripples, you will notice that where the ripples overlap there are quite complicated patterns of waves formed.

The essential point is that at those places where the wave-crests from the two sources arrive at the same time, the waves will work together and the water will be very disturbed, but at points where the crest from one source arrives at the same time as the wave trough from the other source, the waves will cancel each other out, and the water will hardly move. You can hear this effect for sound waves by playing a constant note through stereo speakers.

As you move around a room, you will hear quite large variations in the intensity of sound. Of course, reflections from walls complicate the pattern. This large variation in volume is *not* very noticeable when the stereo is playing music, because music is made up of many frequencies, and they change all the time. The different frequencies, or notes, have their quiet spots in the room in different places.

The other point that should be mentioned is that high

frequency tweeter-like sound is much more *directional* than low frequency woofer-like sound. It really doesn't matter where in the room you put a low-frequency woofer—the sound seems to be all around you anyway. On the other hand, it is quite difficult to get a speaker to spread the high notes in all directions. If you listen to a cheap speaker, the high notes are loudest if the speaker is pointing right at you. A lot of effort has gone into designing tweeters, which are small speakers especially designed to broadcast high notes over a wide angle of directions.

Is Light a Wave

Bearing in mind the above minireview of the properties of waves, let us now reconsider the question of whether light consists of a stream of particles or is some kind of wave. The strongest argument for a particle picture is that light travels in straight lines. You can hear around a corner, at least to some extent, but you certainly can't see. Furthermore, no wave-like interference effects are very evident for light. Finally, it was long known, as we have mentioned, that sound waves were compressional waves in air. If light is a wave, just what is waving? It clearly isn't just air, because light reaches us from the sun, and indeed from stars, and we know the air doesn't stretch that far, or the planets would long ago have been slowed down by air resistance.

Despite all these objections, it was established around 1800 that light *is* in fact some kind of wave. The reason this fact had gone undetected for so long was that the wavelength is *really* short, about one fifty-thousandth of an inch. In contrast, the shortest wavelength sound detectable by humans has a wavelength of about half an inch. The fact that light travels in straight lines is in accord with observations on sound that the higher the frequency (and shorter the wavelength) the greater the tendency to go in straight lines. Similarly, the interference patterns mentioned above for sound waves or ripples on a pond vary over distances of the same sort of size as the wavelengths involved. Patterns like that would not normally be noticeable for light because they would be on such a tiny scale. In fact, it turns out, there *are* ways to see interference effects with light. A familiar example is the many colors often visible in a soap bubble.

These come about because looking at a soap bubble you see light reflected from both sides of a very thin film of water—a thickness that turns out to be comparable to the wavelength of light. The light reflected from the lower layer has to go a little further to reach your eye, so that light wave must wave an extra time or two before getting to your eye compared with the light reflected from the top layer. What you actually *see* is the *sum* of the light reflected from the top layer and that reflected from the bottom layer.

Thinking of this now as the sum of two sets of waves, the light will be bright if the crests of the two waves arrive together, dim if the *crests* of waves reflected from the top layer arrive simultaneously with the *troughs* of waves reflected from the bottom layer. Which of these two possibilities actually occurs for reflection from a particular bit of the soap film depends on just how much further the light reflected from the lower surface has to travel to reach your eye compared with light from the upper surface, and that depends on the angle of reflection and the thickness of the film. Suppose now we shine *white* light on the bubble. White light is made up of all the colors of the rainbow, and these different colors have different wavelengths, so we see colors reflected, because for a particular film, at a particular angle, some colors will be reflected brightly (the crests will arrive together), some dimly, and we will see the ones that win.

If Light is a Wave, What is Waving

Having established that light is a wave, though, we still haven't answered one of the major objections raised above. Just what is waving? We discussed sound waves as waves of compression in air. Actually, that is only one case—sound will also travel through liquids, like water, and solids, like a steel bar. It is found experimentally that, other things being equal, sound travels faster through a medium that is harder to compress: the material just springs back faster and the wave moves through more rapidly. For media of equal springiness, the sound goes faster through the less heavy medium, essentially because the same amount of springiness can push things along faster in a lighter material. So when a sound wave passes, the material—air, water or solid waves as it goes through. Taking this as a hint, it was natural to suppose that light must be just waves in some mysterious material, which was called the *aether*, surrounding and permeating everything. This aether must also fill all of space, out to the stars, because we can see them, so the medium must be there to carry the light. (We could never *hear* an explosion on the moon, however loud, because there is no air to carry the sound to us).

Let us think a bit about what properties this aether must have. Since light travels so fast, it must be very light, and very hard to compress. Yet, as mentioned above, it must allow solid bodies to pass through it freely, without aether resistance, or the planets would be slowing down. Thus we can picture it as a kind of ghostly wind blowing through the earth. But how can we prove any of this? Can we detect it?

Detecting the Aether Wind: the Michelson-Morley Experiment

Detecting the aether wind was the next challenge Michelson set himself after his triumph in measuring the speed of light so accurately. Naturally, something that allows solid bodies to pass through it freely is a little hard to get a grip on. But Michelson realized that, just as the speed of sound is relative to the air, so the speed of light must be relative to the aether. This must mean, if you could measure the speed of light accurately enough, you could measure the speed of light travelling upwind, and compare it with the speed of light travelling downwind, and the difference of the two measurements should be twice the windspeed. Unfortunately, it wasn't that easy.

All the recent accurate measurements had used light travelling to a distant mirror and coming back, so if there was an aether wind along the direction between the mirrors, it would have opposite effects on the two parts of the measurement, leaving a very small overall effect. There was no technically feasible way to do a one-way determination of the speed of light.

At this point, Michelson had a very clever idea for detecting the aether wind. As he explained to his children, it was based on the following puzzle:

Suppose we have a river of width w (say, 100 feet), and two

swimmers who both swim at the same speed v feet per second (say, 5 feet per second). The river is flowing at a steady rate, say 3 feet per second. The swimmers race in the following way: they both start at the same point on one bank. One swims directly across the river to the closest point on the opposite bank, then turns around and swims back. The other stays on one side of the river, swimming upstream a distance (measured along the bank) exactly equal to the width of the river, then swims back to t*he start. Who* wins? Let's consider first the swimmer going upstream and back. Going 100 feet upstream, the speed relative to the bank is only 2 feet per second, so that takes 50 seconds. Coming back, the speed is 8 feet per second, so it takes 12.5 seconds, for a total time of 62.5 seconds.

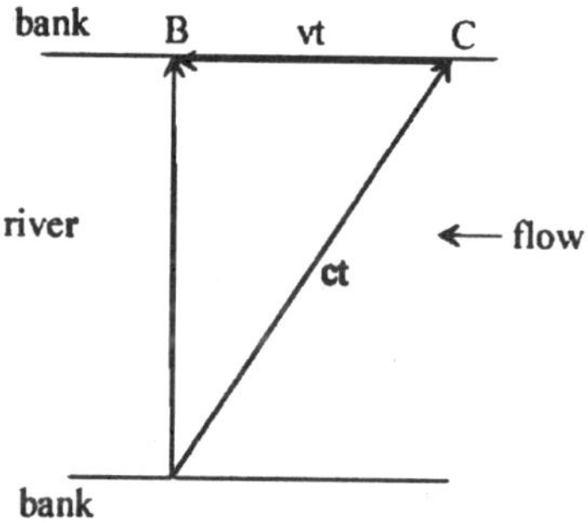

Fig. In time t, the swimmer has moved ct relative to the water, and been carried downstream a distance vt.

The swimmer going across the flow is trickier. It won't do simply to aim directly for the opposite bank: the flow will carry the swimmer downstream. To succeed in going directly across, the swimmer must actually aim upstream at the correct angle. Thus, the swimmer is going at 5 feet per second, at an angle, relative to the river, and being carried downstream at a rate of 3 feet per second. If the angle is correctly chosen so that the net movement is directly across, in one second the swimmer must have moved *four feet* across: the distances covered in one second will form a 3, 4, 5 triangle. So, at a crossing rate of 4 feet per second, the swimmer gets across in 25 seconds, and back in the same time, for a total time of 50 seconds. The cross-stream swimmer wins.

This turns out to true whatever their swimming speed. The source of light is at *s*, the 45 degree line is the half-silvered mirror, *b* and *c* are mirrors and d the observer. Michelson's great idea was to construct an exactly similar race for pulses of light, with the aether wind playing the part of the river. The scheme of the experiment is as follows: a pulse of light is directed at an angle of 45 degrees at a half-silvered, half transparent mirror, so that half the pulse goes on through the glass, half is reflected. These two half-pulses are the two swimmers.

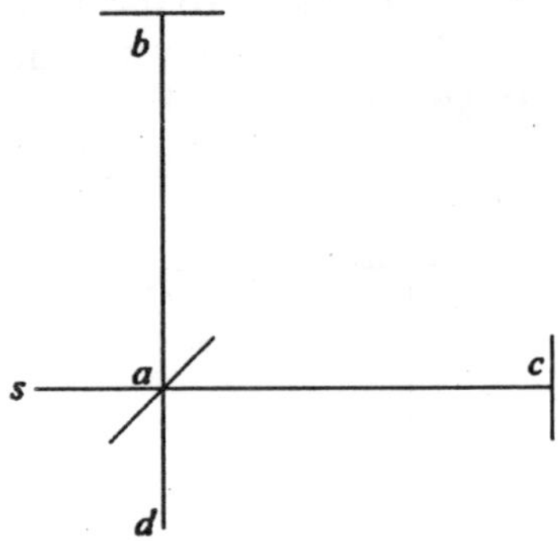

Fig. This diagram is from the original paper.

They both go on to distant mirrors which reflect them back to the half-silvered mirror. At this point, they are again half reflected and half transmitted, but a telescope is placed behind the half-silvered mirror as shown in the figure so that half of each half-pulse will arrive in this telescope.

Now, if there is an aether wind blowing, someone looking through the telescope should see the halves of the two half-pulses to arrive at slightly different times, since one would have gone more upstream and back, one more across stream in general. To maximize the effect, the whole apparatus, including the distant mirrors, was placed on a large turntable so it could be swung around.

Let us think about what kind of time delay we expect to find between the arrival of the two half-pulses of light. Taking the speed of light to be *c* miles per second relative to the aether, and the aether to be flowing at *v* miles per second through the laboratory, to go a distance *w* miles upstream will take $w/(c-v)$ seconds, then to come back will take $w/(c + v)$ seconds. The total roundtrip time upstream and downstream is the sum of these, which works out to be $2wc/(c^2-v^2)$, which can also be written $(2w/c) \times 1/(1-v^2/c^2)$. Now, we can safely assume the speed of the aether is much less than the speed of light, otherwise it would have been noticed long ago, for example in timing of eclipses of Jupiter's satellites.

This means v^2/c^2 is a very small number, and we can use some handy mathematical facts to make the algebra a bit easier. First, if x is very small compared to 1, $1/(1-x)$ is very close to $1+x$. (You can check it with your calculator). Another fact we shall need in a minute is that for small x, the square root of $1+x$ is very close to $1+x/2$.

Putting all this together,

$$\text{upstream-downstream roundtrip time} \cong \frac{2w}{c} \times \left(1+\frac{v^2}{c^2}\right)$$

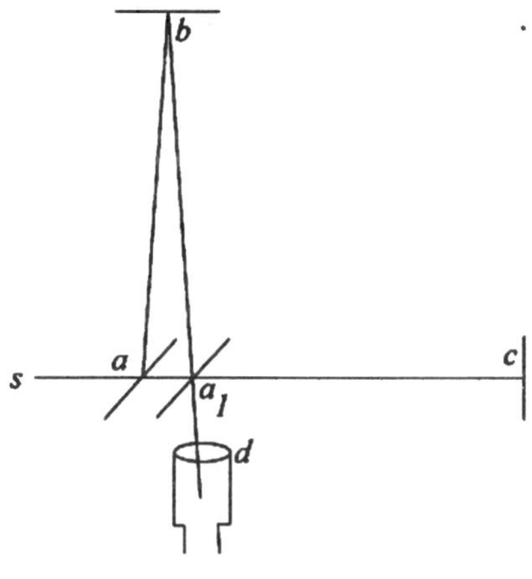

Fig. This is also from the original paper, and shows the expected path of light relative to the aether with an aether wind blowing

Now, what about the cross-stream time? The actual cross-stream speed must be figured out as in the example above using a right-angled triangle, with the hypoteneuse equal to the speed c, the shortest side the aether flow speed v, and the other side the cross-stream speed we need to find the time to get across. From Pythagoras' theorem, then, the cross-stream speed is the square root of (c^2-v^2).

Since this will be the same both ways, the roundtrip cross-stream time will be

$$2w/\sqrt{c^2-v^2}.$$

This can be written in the form

$$\frac{2w}{c}\frac{1}{\sqrt{1-v^2/c^2}} \cong \frac{2w}{c}\frac{1}{1-\left(\frac{1}{v^2/2c^2}\right)} \cong \frac{2w}{c}\left(1+\frac{v^2}{2c^2}\right).$$

where the two successive approximations, valid for $v/c = x << 1$, are $\sqrt{1-x} \cong 1-(x)$ and $1/(1-x) \cong 1+x$.

Therefore the

$$\text{cross-stream roundtrip time} \cong \frac{2w}{c} \times \left(1+\frac{v^2}{2c^2}\right).$$

Looking at the two roundtrip times at the ends of the two paragraphs above, we see that they differ by an amount $(2w/c) \times v^2/2c^2$. Now, $2w/c$ is just the time the light would take if there were no aether wind at all, say, a few millionths of a second. If we take the aether windspeed to be equal to the earth's speed in orbit, for example, v/c is about 1/10, 000, so v^2/c^2 is about 1/100, 000, 000.

This means the time delay between the pulses reflected from the different mirrors reaching the telescope is about one-hundred-millionth of a few millionths of a second. It seems completely hopeless that such a short time delay could be detected. However, this turns out not to be the case, and Michelson was the first to figure out how to do it. The trick is to use the interference properties of the lightwaves. Instead of sending pulses of light, Michelson sent in a steady beam of light of a single colour.

This can be visualized as a sequence of ingoing waves, with a wavelength one fifty-thousandth of an inch or so. Now this sequence of waves is split into two, and reflected as previously described. One set of waves goes upstream and downstream, the other goes across stream and back. Finally, they come together into the telescope and the eye.

If the one that took longer is half a wavelength behind, its troughs will be on top of the crests of the first wave, they will cancel, and nothing will be seen. If the delay is less than that, there will still be some dimming. However, slight errors in the placement of the mirrors would have the same effect. This is one reason why the apparatus is built to be rotated. On turning it through 90 degrees, the upstream-downstream and the cross-stream waves change places. Now the other one should be behind. Thus, if there is an aether wind, if you watch through the telescope while you rotate the turntable, you should expect to see variations in the brightness of the incoming light.

To magnify the time difference between the two paths, in the actual experiment the light was reflected backwards and forwards several times, like a several lap race.

Michelson calculated that an aether windspeed of only one or two miles a second would have observable effects in this experiment, so if the aether windspeed was comparable to the earth's speed in orbit around the sun, it would be easy to see. In fact, *nothing* was observed. The light intensity did not vary at all. Some time later, the experiment was redesigned so that an aether wind caused by the earth's daily rotation could be detected. Again, nothing was seen.

Finally, Michelson wondered if the aether was somehow getting stuck to the earth, like the air in a below-decks cabin on a ship, so he redid the experiment on top of a high mountain in California. Again, no aether wind was observed. It was difficult to believe that the aether in the immediate vicinity of the earth was stuck to it and moving with it, because light rays from stars would deflect as they went from the moving faraway aether to the local stuck aether.

The only possible conclusion from this series of very difficult experiments was that the whole concept of an all-pervading aether was wrong from the start. Michelson was very reluctant to think along these lines. In fact, new theoretical insight into the nature of light had arisen in the 1860's from the brilliant theoretical work of Maxwell, who had written down a set of equations describing how electric and magnetic fields can give rise to each other.

He had discovered that his equations predicted there could be waves made up of electric and magnetic fields, and the speed of these waves, deduced from experiments on how these fields link together, would be 186, 300 miles per second. This is, of course, the speed of light, so it is natural to assume that light is made up of fast-varying electric and magnetic fields. But this leads to a big problem: Maxwell's equations predict a definite speed for light, and it *is* the speed found by measurements.

But what is the speed to be measured relative to? The whole point of bringing in the aether was to give a picture for light resembling the one we understand for sound, compressional waves in a medium. The speed of sound through air is measured relative to air. If the wind is blowing towards you from the source of sound, you will hear the sound sooner. If there isn't an aether, though, this analogy doesn't hold up. So what does light travel at 186, 300 miles per second relative to?

There is another obvious possibility, which is called the emitter theory: the light travels at 186, 300 miles per second relative to the source of the light. The analogy here is between light emitted by a source and bullets emitted by a machine gun. The bullets come out at a definite speed (called the muzzle velocity) relative to the barrel of the gun.

If the gun is mounted on the front of a tank, which is moving forward, and the gun is pointing forward, then relative to the ground the bullets are moving faster than they would if shot from a tank at rest. The simplest way to test the emitter theory of light, then, is to measure the speed of light emitted in the forward direction by a flashlight moving in the forward direction, and see if it exceeds the known speed of light by an amount equal to the speed of the flashlight.

Actually, this kind of direct test of the emitter theory only became experimentally feasible in the nineteen-sixties. It is now possible to produce particles, called neutral pions, which decay each one in a little explosion, emitting a flash of light. It is also possible to have these pions moving forward at 185, 000 miles per second when they self destruct, and to catch the light emitted in the forward direction, and clock its speed.

It is found that, despite the expected boost from being emitted by a very fast source, the light from the little explosions is going forward at the usual speed of 186, 300 miles per second. In the last century, the emitter theory was rejected because it was thought the appearance of certain astronomical phenomena, such as double stars, where two stars rotate around each other, would be affected. Those arguments have since been criticized, but the pion test is unambiguous.

Einstein's Answer

The results of the various experiments discussed above seem to leave us really stuck. Apparently light is not like sound, with a definite speed relative to some underlying medium. However, it is also not like bullets, with a definite speed relative to the source of the light. Yet when we measure its speed we always get the same result. How can all these facts be interpreted in a simple consistent way?

SPECIAL RELATIVITY

Galilean Relativity Again

At this point in the course, we finally enter the twentieth century—Albert Einstein wrote his first paper on relativity in 1905. To put his work in context, let us first review just what is meant by "relativity" in physics. The first example, mentioned in a previous lecture, is what is called "Galilean relativity" and is nothing but Galileo's perception that by observing the motion of objects, alive or dead, in a closed room there is no way to tell if the room is at rest or is in fact in a boat moving at a steady speed in a fixed direction. (You *can* tell if the room is accelerating or turning around).

Everything looks the same in a room in steady motion as it does in a room at rest. After Newton formulated his Laws of Motion, describing how bodies move in response to forces and so on, physicists reformulated Galileo's observation in a slightly more technical, but equivalent, way: they said the laws of physics are the same in a uniformly moving room as they are in a room at rest.

In other words, the same force produces the same acceleration, and an object experiencing no force moves at a steady speed in a straight line in either case. Of course, talking in these terms implies that we have clocks and rulers available so that we can actually time the motion of a body over a measured distance, so the physicist envisions the room in question to have calibrations along all the walls, so the position of anything can be measured, and a good clock to time motion.

Such a suitably equipped room is called a "frame of reference"—the calibrations on the walls are seen as a frame which you can use to specify the precise position of an object at a given time. (This is the same as a set of "coordinates"). Anyway, the bottom line is that no amount of measuring of motions of objects in the "frame of reference" will tell you whether this is a frame at rest or one moving at a steady velocity.

What exactly do we mean by a frame "at rest" anyway? This seems obvious from our perspective as creatures who live on the surface of the earth—we mean, of course, at rest relative to fixed objects on the earth's surface. Actually, the earth's rotation means this isn't quite a fixed frame, and also the earth is moving in orbit at 18 miles per second. From an astronaut's point of view, then, a frame fixed relative to the sun might seem more reasonable.

But why stop there? We believe the laws of physics are good throughout the universe. Let us consider somewhere in space far from the sun, even far from our galaxy. We would see galaxies in all directions, all moving in different ways. Suppose we now set up a frame of reference and check that Newton's laws still work.

In particular, we check that the First Law holds—that a body experiencing no force moves at a steady speed in a straight line. This First law is often referred to as The Principle of Inertia, and a frame in which it holds is called an Inertial Frame. Then we set up another frame of reference, moving at a steady velocity relative to the first one, and find that Newton's laws are o.k. in this frame too.

The point to notice here is that it is not at all obvious which—if either—of these frames is "at rest". We *can*, however, assert that they are both *inertial* frames, after we've checked that in both of them, a body with no forces acting on it moves at a steady speed in a straight line (the speed could be zero). In this situation, Michelson would have said that a frame "at rest" is one at rest relative to the aether. However, his own experiment found motion through the aether to be undetectable, so how would we ever know we were in the right frame?

The middle of the nineteenth century there was a substantial advance in the understanding of electric and magnetic fields. (In fact, this advance is in large part responsible for the improvement in living standards since that time). The new understanding was summarized in a set of equations called Maxwell's equations describing how electric and magnetic fields interact and give rise to each other, just as, two centuries earlier, the new understanding of dynamics was summarized in the set of equations called Newton's laws.

The important thing about Maxwell's equations for our present purposes is that they predicted waves made up of electric and magnetic fields that moved at $3X10^8$ meters per second, and it was immediately realized that this was no coincidence—light waves must be nothing but waving electric and magnetic fields. (This is now fully established to be the case).

It is worth emphasizing that Maxwell's work predicted the speed of light from the results of experiments that were not thought at the time they were done to have anything to do with light—experiments on, for example, the strength of electric field produced by waving a magnet. Maxwell was able to deduce a speed for waves like this using methods analogous to those by which earlier scientists had figured out the speed of sound from a knowledge of the density and the springiness of air.

Generalizing Galilean Relativity to Include Light: Special Relativity

We now come to Einstein's major insight: the Theory of Special Relativity. It is deceptively simple. Einstein first dusted off Galileo's discussion of experiments below decks on a uniformly moving ship, and restated it as: The Laws of Physics are the same in all Inertial Frames. Einstein then simply brought this up to date, by pointing out that the Laws of Physics must now include Maxwell's equations describing electric and magnetic fields as well as Newton's laws describing motion of masses under gravity and other forces. (Note for experts and the curious: we shall find that Maxwell's equations are completely unaltered by special relativity, but, as will become clear later, Newton's Laws do need a bit of readjustment to include special relativistic phenomena.

The First Law is still o.k., the Second Law in the form $F = ma$ is not, because we shall find mass varies; we need to equate force to rate of change of momentum (Newton understood that, of course—that's the way he stated the law!). The Third Law, stated as action equals reaction, no longer holds because if a body moves, its electric field, say, does not readjust instantaneously—a ripple travels outwards at the speed of light.

Before the ripple reaches another charged body, the electric forces between the two will be unbalanced. However, the crucial consequence of the Third Law—the conservation of momentum when two bodies interact, still holds. It turns out that the rippling field itself carries momentum, and everything balances).

Demanding that Maxwell's equations be satisfied in all inertial frames has one major consequence as far as we are concerned. Maxwell's equations give the speed of light to be $3X10^8$ meters per second.

Therefore, demanding that the laws of physics are the same in all inertial frames implies that the speed of any light wave, measured in any inertial frame, must be 3×10^8 meters per second. This then is the entire content of the Theory of Special Relativity: the Laws of Physics are the same in any inertial frame, and, in particular, any measurement of the speed of light in any inertial frame will always give $3X10^8$ meters per second.

You *Really* Can't Tell You're Moving

Just as Galileo had asserted that observing gnats, fish and dripping bottles, throwing things and generally jumping around would not help you to find out if you were in a room at rest or moving at a steady velocity, Einstein added that no kind of observation at all, even measuring the speed of light across your room to any accuracy you like, would help find out if your room was "really at rest". This implies, of course, that the concept of being "at rest" is meaningless. If Einstein is right, there is *no* natural rest-frame in the universe. Naturally, there can be no "aether", no thin transparent jelly filling space and vibrating with light waves, because if there were, *it* would provide the natural rest frame, and affect the speed of light as measured in other moving inertial frames.

So we see the Michelson-Morley experiment was doomed from the start. There never was an aether wind. The light was not slowed down by going "upstream"—light *always* travels at the same speed, which we shall now call c,

$$c = 3 \times 10^8 \text{ meters per second}$$

to save writing it out every time. This now answers the question of what the speed of light, c, is relative to. We already found that it is not like sound, relative to some underlying medium.

It is also not like bullets, relative to the source of the light (the discredited emitter theory). Light travels at c relative to the observer, since if the observer sets up an inertial frame (clocks, rulers, etc). to measure the speed of light he will find it to be c. (We always assume our observers are very competent experimentalists!)

Truth and Consequences

The Truth we are referring to here is the seemingly innocuous and plausible sounding statement that all inertial frames are as good as each other—the laws of physics are the same in all of them—and so the *speed of light* is the same in all of them. This Special Theory of Relativity has some surprising consequences, which reveal themselves most dramatically when things are moving at relative speeds comparable to the speed of light.

Einstein liked to explain his theory using what he called "thought experiments" involving trains and other kinds of transportation moving at these speeds (technically unachievable so far!), and we shall follow his general approach.

To begin with, let us consider a simple measurement of the speed of light carried out at the same time in two inertial frames moving at half the speed of light relative to each other. The setup is as follows: on a flat piece of ground, we have a flashlight which emits a blip of light, like a strobe. We have two photocells, devices which click and send a message down a wire when light falls on them.

The photocells are placed 10 meters apart in the path of the blip of light, they are somehow wired into a clock so that the time taken by the blip of light to travel from the first photocell to the second, in other words, the time between clicks, can be measured. From this time and the known distance between them, we can easily find the speed of the blip of light.

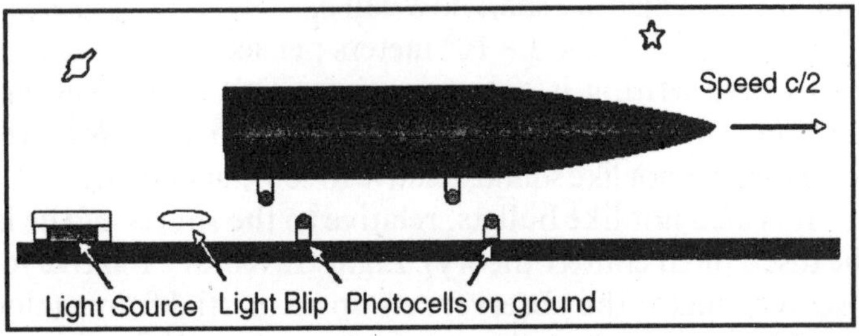

The speed of the same blip of light is measured by two observers, having relative speed c/2. Both measure the time the blip takes from the one photocell to a second one 10 meters further on. Both find the speed to be c. Meanwhile, there is another observer, passing overhead in a spaceship traveling at half the speed of light. She is also equipped with a couple of photocells, placed 10 meters apart on the bottom of her spaceship as shown, and she is able to measure the speed of the same blip of light, relative to her frame of reference (the spaceship).

The observer on the spaceship will measure the blip of light to be traveling at *c* relative to the spaceship, the observer on the ground will measure the same blip to be traveling at *c* relative to the ground. That is the unavoidable consequence of the Theory of Relativity.

Special Relativity in a Nutshell

Einstein's Theory of Special Relativity, may be summarized as follows: The Laws of Physics are the same in any Inertial Frame of Reference. (Such frames move at steady velocities with respect to each other).

These Laws include in particular Maxwell's Equations describing electric and magnetic fields, which predict that light always travels at a particular speed c, equal to about 3×10^8 meters per second, that is, 186, 300 miles per second. It follows that any measurement of the speed of any flash of light by any observer in any inertial frame will give the same answer c.

We have already noted one counter-intuitive consequence of this, that two different observers moving relative to each other, each measuring the speed of the *same* blob of light relative to himself, will *both* get *c*, even if their relative motion is in the same direction as the motion of the blob of light. We shall now explore how this simple assumption changes everything we thought we understood about time and space.

A Simple but Reliable Clock

We mentioned earlier that each of our (inertial) frames of reference is calibrated (had marks at regular intervals along the walls) to measure distances, and has a clock to measure time. Let us now get more specific about the clock—we want one that is easy to understand in any frame of reference. Instead of a pendulum swinging back and forth, which wouldn't work away from the earth's surface anyway, we have a blip of light bouncing back and forth between two mirrors facing each other.

We call this device a *light clock*. To really use it as a timing device we need some way to count the bounces, so we position a photocell at the upper mirror, so that it catches the edge of the blip of light.

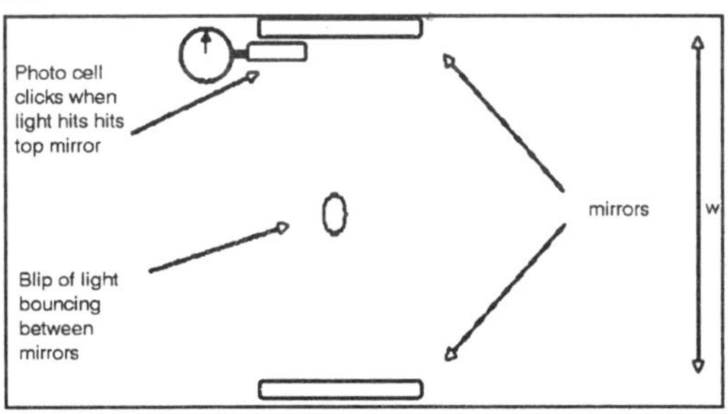

The photocell clicks when the light hits it, and this regular series of clicks drives the clock hand around, just as for an ordinary clock. Of course, driving the photocell will eventually use up the blip of light, so we also need some provision to reinforce the blip occasionally, such as a strobe light set to flash just as it passes and thus add to the intensity of the light.

Admittedly, this may not be an easy way to build a clock, but the basic idea is simple.

Einstein's light Clock: a blip of light bounces between two parallel mirrors w meters apart; each time it hits the top mirror, the photocell moves the clock on one click. Time between clicks: 2w/c.

It's easy to figure out how frequently our light clock clicks. If the two mirrors are a distance w apart, the round trip distance for the blip from the photocell mirror to the other mirror and back is $2w$. Since we know the blip always travels at c, we find the round trip time to be $2w/c$, so this is the time between clicks.

This isn't a very long time for a reasonable sized clock! The crystal in a quartz watch "clicks " of the order of 10, 000 times a second. That would correspond to mirrors about nine miles apart, so we need our clock to click about 1, 000 times faster than that to get to a reasonable size. Anyway, let us assume that such purely technical problems have been solved.

Looking at Somebody Else's Clock

Let us now consider two observers, Jack and Jill, each equipped with a calibrated inertial frame of reference, and a light clock. To be specific, imagine Jack standing on the ground with his light clock next to a straight railroad line, while Jill and her clock are on a large flatbed railroad wagon which is moving down the track at a constant speed v.

Jack now decides to check Jill's light clock against his own. He knows the time for his clock is $2w/c$ between clicks. Imagine it to be a slightly misty day, so with binoculars he can actually see the blip of light bouncing between the mirrors of Jill's clock.

How long does he think that blip takes to make a round trip? The one thing he's sure of is that it must be moving at c = 186, 300 miles per second, relative to him—that's what Einstein tells him. So to find the round trip time, all he needs is the round trip distance. This will *not* be $2w$, because the mirrors are on the flatbed wagon moving down the track, so, relative to Jack on the ground, when the blip gets back to the top mirror, that mirror has moved down the track some since the blip left, so the blip actually follows a zigzag path as seen from the ground.

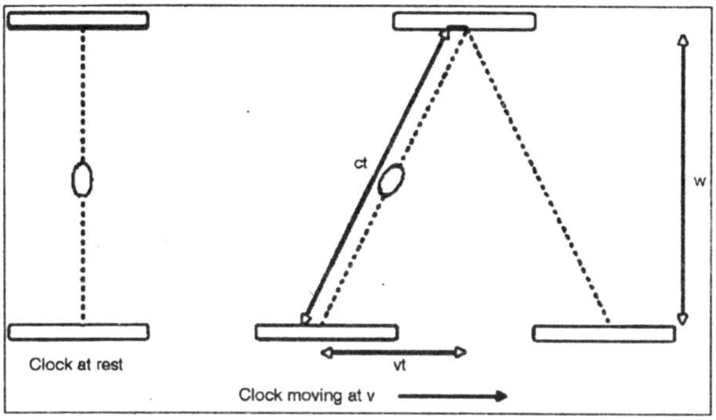

Two identical light clocks: one at rest, one moving relative to us. The light blips in both travel at the same speed relative to us, the one in the moving clock goes further, so must take longer between clicks, t is the time from one mirror to the other.

Suppose now the blip in Jill's clock on the moving flatbed wagon takes time t to get from the bottom mirror to the top mirror as measured by Jack standing by the track. Then the length of the "zig" from the bottom mirror to the top mirror is necessarily ct, since that is the distance covered by any blip of light in time t.

Meanwhile, the wagon has moved down the track a distance vt, where v is the speed of the wagon. This should begin to look familiar it is precisely the same as the problem of the swimmer who swims at speed c relative to the water crossing a river flowing at v! We have again a right-angled triangle with hypotenuse ct, and shorter sides vt and w.

From Pythagoras, then,

$$c^2t^2 = v^2t^2 + w^2$$

so

$$t^2(c^2 - v^2) = w^2$$

or

$$t^2(1 - v^2/c^2) = w^2/c^2$$

and, taking the square root of each side, then doubling to get the round trip time, we conclude that Jack sees the time between clicks for Jill's clock to be:

$$\text{time between clicks for moving clock} = \frac{2w}{c}\frac{1}{\sqrt{1-v^2/c^2}}.$$

Of course, this gives the right answer $2w/c$ for a clock at rest, that is, $v = 0$.

This means that Jack sees Jill's light clock to be going slow—a longer time between clicks—compared to his own identical clock. Obviously, the effect is not dramatic at real railroad speeds. The correction factor is $\sqrt{1-v^2/c^2}$, which differs from 1 by about one part in a trillion even for a bullet train! Nevertheless, the effect is real and can be measured.

It is important to realise that the only reason we chose a *light* clock, as opposed to some other kind of clock, is that its motion is very easy to analyse from a different frame. Jill could have a collection of clocks on the wagon, and would synchronize them all. For example, she could hang her wristwatch right next to the face of the light clock, and observe them together to be sure they always showed the same time. Remember, in her frame her light clock clicks every $2w/c$ seconds, as it is designed to do.

Observing this scene from his position beside the track, Jack will see the synchronized light clock and wristwatch next to each other, and, of course, note that the wristwatch is *also* running slow by the factor $\sqrt{1-v^2/c^2}$. In fact, *all* her clocks, including her pulse, are slowed down by this factor according to Jack. Jill is aging more slowly because she's moving! But this isn't the whole story—we must now turn everything around and look at it from Jill's point of view. Her inertial frame of reference is just as good as Jack's. She sees his light clock to be moving at speed v (backwards) so from her point of view his light blip takes the longer zigzag path, which means his clock runs slower than hers.

That is to say, each of them will see the other to have slower clocks, and be aging more slowly. This phenomenon is called time dilation. It has been verified in recent years by flying very accurate clocks around the world on jetliners and finding they register less time, by the predicted amount, than identical clocks left on the ground. Time dilation is also very easy to observe in elementary particle physics.

Fitzgerald Contraction

Consider now the following puzzle: suppose Jill's clock is

equipped with a device that stamps a notch on the track once a second. How far apart are the notches? From Jill's point of view, this is pretty easy to answer. She sees the track passing under the wagon at v meters per second, so the notches will of course be v meters apart. But Jack sees things differently. He sees Jill's clocks to be running slow, so he will see the notches to be stamped on the track at intervals of $1/\sqrt{1-v^2/c^2}$ seconds (so for a relativistic train going at $v = 0.8c$, the notches are stamped at intervals of $5/3 = 1.67$ seconds). Since Jack agrees with Jill that the relative speed of the wagon and the track is v, he will assert the notches are not v meters apart, but $v/\sqrt{1-v^2/c^2}$ meters apart, a greater distance.

Who is right? It turns out that Jack is right, because the notches are in his frame of reference, so he can wander over to them with a tape measure or whatever, and check the distance. This implies that as a result of her motion, Jill observes the notches to be closer together by a factor $\sqrt{1-v^2/c^2}$ than they would be at rest. This is called the *Fitzgerald contraction*, and applies not just to the notches, but also to the track and to Jack—everything looks somewhat squashed in the direction of motion!

Experimental Evidence for Time Dilation: Dying Muons

The first clear example of time dilation was provided over fifty years ago by an experiment detecting *muons*. These particles are produced at the outer edge of our atmosphere by incoming cosmic rays hitting the first traces of air.

They are unstable particles, with a "half-life" of 1.5 microseconds (1.5 millionths of a second), which means that if at a given time you have 100 of them, 1.5 microseconds later you will have about 50, 1.5 microseconds after that 25, and so on. Anyway, they are constantly being produced many miles up, and there is a constant rain of them towards the surface of the earth, moving at very close to the speed of light. In 1941, a detector placed near the top of Mount

Washington (at 6000 feet above sea level) measured about

570 muons per hour coming in. Now these muons are raining down from above, but dying as they fall, so if we move the detector to a lower altitude we expect it to detect fewer muons because a fraction of those that came down past the 6000 foot level will die before they get to a lower altitude detector. Approximating their speed by that of light, they are raining down at 186, 300 miles per second, which turns out to be, conveniently, about 1, 000 feet per microsecond.

Thus they should reach the 4500 foot level 1.5 microseconds after passing the 6000 foot level, so, if half of them die off in 1.5 microseconds, as claimed above, we should only expect to register about 570/2 = 285 per hour with the same detector at this level. Dropping another 1500 feet, to the 3000 foot level, we expect about 280/2 = 140 per hour, at 1500 feet about 70 per hour, and at ground level about 35 per hour.

To summarize: given the known rate at which these raining-down unstable muons decay, and given that 570 per hour hit a detector near the top of Mount Washington, we only expect about 35 per hour to survive down to sea level. In fact, when the detector was brought down to sea level, it detected about 400 per hour! How did they survive? The reason they didn't decay is that in their frame of reference, much less *time had passed.*

Their actual speed is about $0.994c$, corresponding to a time dilation factor of about 9, so in the 6 microsecond trip from the top of Mount Washington to sea level, their clocks register only 6/9 = 0.67 microseconds. In this period of time, only about one-quarter of them decay.

What does this look like from the muon's point of view? How do they manage to get so far in so little time? To them, Mount Washington and the earth's surface are approaching at 0.994c, or about 1, 000 feet per microsecond. But in the 0.67 microseconds it takes them to get to sea level, it would seem that to them sea level could only get 670 feet closer, so how could they travel the whole 6000 feet from the top of Mount Washington?

The answer is the Fitzgerald contraction. To them, Mount Washington is squashed in a vertical direction (the direction of motion) by a factor of $\sqrt{1-v^2/c^2}$,the same as the time dilation factor, which for the muons is about 9. So, to the muons, Mount Washington is only 670 feet high—this is why they can get down it so fast.

SYNCHRONIZING CLOCKS

Suppose we want to synchronize two clocks that are some distance apart. We could stand beside one of them and look at the other through a telescope, but we'd have to remember in that case that we are seeing the clock *as it was when the light left it*, and correct accordingly.

Another way to be sure the clocks are synchronized, assuming they are both accurate, is to start them together. How can we do that? We could, for example, attach a photocell to each clock, so when a flash of light reaches the clock, it begins running. If, then, we place a flashbulb at the midpoint of the line joining the two clocks, and flash it, the light flash will take the same time to reach the two clocks, so they will start at the same time, and therefore be synchronized.

Let us now put this whole arrangement - the two clocks and the midpoint flashbulb - on a train, and we suppose the train is moving at some speed v to the right, say half the speed of light or so.

Let's look carefully at the clock-synchronizing operation as seen from the ground. In fact, an observer on the ground would say the clocks are *not* synchronized by this operation! The basic reason is that he would see the flash of light from the middle of the train traveling at *c relative to the ground* in each direction, but he would also observe the back of the train coming at v to meet the flash, whereas the front is moving at v away from the bulb, so the light flash must go further to catch up.

In fact, it is not difficult to figure out how much later the flash reaches the front of the train compared with the back of the train, as viewed from the ground. First recall that as viewed from the ground the train has length $L\sqrt{1-v^2/c^2}$

The train is moving to the right: the central bulb emits a flash of light. Seen from the ground, the part of the flash moving towards the rear travels at c, the rear travels at v to meet it. Letting t_B be the time it takes the flash to reach the back of the train, it is clear from the figure that from which t_B is given by In a similar way, the time for the flash of light to reach the front of the train.

Therefore the time difference between the starting of the two clocks, as seen from the ground, is. Remember, this is the time difference between the starting of the train's back clock and its front clock as measured by an observer on the ground with clocks on the ground. However, to this observer the clocks on the train appear to tick more slowly, by the factor, so that although the ground observer measures the time interval between the starting of the clock at the back of the train and the clock at the front as seconds, he also sees the slow running clock at the back actually reading seconds at the instant he sees the front clock to start.

To summarize: as seen from the ground, the two clocks on the train (which is moving at v in the x-direction) are running slowly, registering only seconds for each second that passes. Equally important, the clocks—which are synchronized by an observer on the train—appear unsynchronized when viewed from the ground, the one at the back of the train reading seconds ahead of the clock at the front of the train, where L is the rest length of the train (the length as measured by an observer on the train).

Note that if $L = 0$, that is, if the clocks are together, both the observers on the train and those on the ground will agree that they are synchronized. We need a *distance* between the clocks, as well as relative motion, to get a disagreement about synchronization.

RELATIVITY CONNECTS ELECTRIC AND MAGNETIC FIELDS

Suppose we have an infinitely long straight wire, having a charge density of electrons of $-\lambda$ coulombs per meter, all moving at speed v to the right (recall typical speeds are centimeters per minute) and a neutralizing fixed background of positive charge, also of course λ coulombs per meter. The current in the wire has magnitude $I = \lambda v$ (and actually is flowing to the left, since the moving electrons carry negative charge).

Suppose also that a positive charge q is outside the wire, a distance r from the axis, and this outside charge is moving at the same exact velocity as the electrons in the wire.

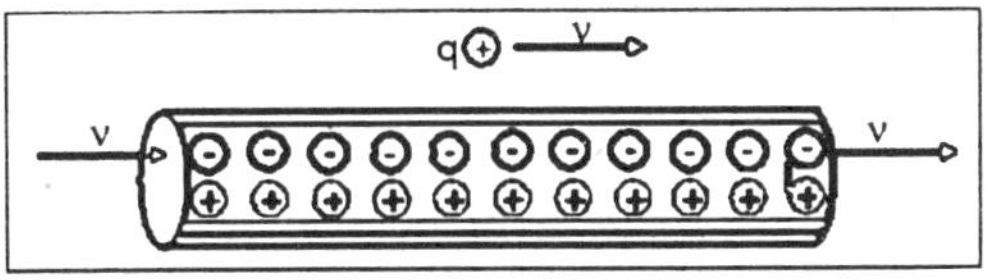

Positive charges (ions) are fixed: negative charges (electrons) drift at speed v. Positive charge outside wire also moves at v.

Note: Positive and negative charges are of course uniformly distributed throughout the wire. They are shown separated here for clarity.

What force does the positive charge q feel?

The wire is electrically neutral, since it contains equal densities of positive and negative charges, both uniformly distributed throughout the wire (the illustration above is of course schematic). So *q* feels no electrical force.

However, since *q* is moving, it will feel a magnetic force,

$$\vec{F}_{mag} = q\vec{v} \times \vec{B}.$$

The magnetic field lines are vertical circles centered along the axis of the wire, the field being into the page where the charge is and of magnitude

$$B = \frac{\mu_0 I}{2\pi r} = \frac{\mu_0 \lambda v}{2\pi r}$$

so the force on the charge is of magnitude

$$F = \frac{q\mu_0 \lambda v^2}{2\pi r}$$

and is directed away from the wire, so the charge will accelerate away from the wire. Now let us examine the same physical system in the frame of reference in which the charge is initially at rest. In that frame, the electrons are also at rest, but the positive background charge is flowing at *v*:

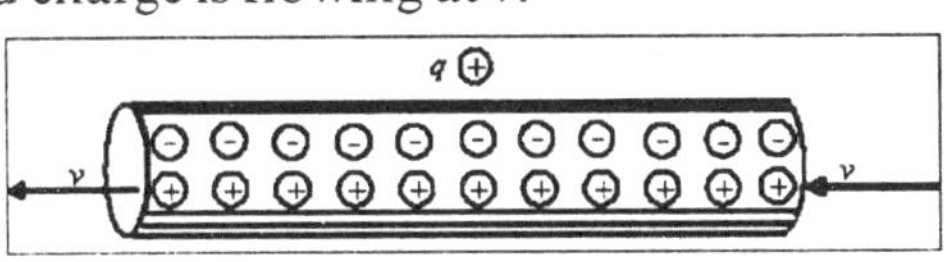

The view in the rest frame of the electrons and the external postive charge.

What force does q feel in this frame?

Since q is at rest, it cannot feel a magnetic force: such forces depend linearly on speed!

Yet it looks as if it can't feel an electric force either, because the positive and negative charges in the wire have equal densities, right? This leads to the conclusion that q feels no force at all in this frame, so it won't accelerate away from the wire, as it did in the other frame. This is of course nonsense—so where did we go wrong?

Solved Electrically

The mistake was in ignoring the relativistic Fitzgerald-Lorentz contraction, even though the velocities involved are *millimeters per second*! In the frame in which the wire is at rest, the positive and negative charge densities exactly balance, otherwise there will be extra electrostatic fields that the electrons will quickly move to neutralize.

However, this *necessarily* means that the densities cannot balance exactly in the frame in which the drifting electrons are at rest. In the electrons' rest frame, the positive charges, which had density λ coulombs per meter in their rest frame, are moving at speed v, so relativistic contraction will increase their density to:

$$\frac{\lambda}{\sqrt{1-v^2/c^2}} \cong \lambda + \frac{\lambda v^2}{2c^2}.$$

On the other hand, in this electron frame the electrons are at rest, so their density is actually less than it was in the frame of the wire, it has decreased by $\lambda v^2/2c^2$. The net effect is that the wire, electrically neutral in the lab frame, has a positive charge density $\lambda v^2/c^2$ in the frame of the moving electrons (and the outside charge). Recall Gauss's Law relating the total electric field flux through a surface to the enclosed charge,

$$\int \vec{E}.d\vec{A} = (\text{enclosed charge})/\varepsilon_0.$$

Exercise: use this to verify that the electric field from an infinite line of charge λ coulombs per meter has magnitude $\lambda/2\pi r\varepsilon_0$ at distance r from the line charge From the exercise result, the electrostatic force on the charge q is:

$$F = qE = q\lambda \frac{v^2}{c^2}\frac{1}{2\pi r\varepsilon_0} = \frac{q\lambda v^2}{2\pi r}\frac{1}{\varepsilon_0 c^2} = \frac{q\lambda v^2}{2\pi r}\mu_0$$

where in the last step we used $1/c^2 = \mu_0\varepsilon_0$.

This purely electrical force is identical in magnitude to the purely magnetic force in the other frame! So observers in the two frames *will* agree on the rate at which the particle accelerates away from the wire, but one will call the accelerating force magnetic, the other electric. We are forced to the conclusion that whether a particular force on an actual particle is magnetic or electric, or some mixture of both, depends on the frame of reference—so the distinction is rather artificial.

EINSTEIN'S PARABLE

In Einstein's little book *Relativity: the Special and the General Theory*, he introduces general relativity with a parable. He imagines going into deep space, far away from gravitational fields, where any body moving at steady speed in a straight line will continue in that state for a very long time. He imagines building a space station out there - in his words, "a spacious chest resembling a room with an observer inside who is equipped with apparatus." Einstein points out that there will be no gravity, the observer will tend to float around inside the room.

But now a rope is attached to a hook in the middle of the lid of this "chest" and an unspecified "being" pulls on the rope with a constant force. The chest and its contents, including the observer, accelerate "upwards" at a constant rate. How does all this look to the man in the room? He finds himself moving towards what is now the "floor" and needs to use his leg muscles to stand. If he releases anything, it accelerates towards the floor, and in fact all bodies accelerate at the same rate. If he were a normal human being, he would assume the room to be in a gravitational field, and might wonder why the room itself didn't fall. Just then he would discover the hook and rope, and conclude that the room was suspended by the rope.

Einstein asks: should we just smile at this misguided soul? His answer is no - the observer in the chest's point of view is just as valid as an outsider's.

In other words, being inside the (from an outside perspective) uniformly accelerating room is physically equivalent to being in a uniform gravitational field. This is the basic postulate of general relativity.

Special relativity said that all inertial frames were equivalent. General relativity extends this to accelerating frames, and states their equivalence to frames in which there is a gravitational field. This is called the Equivalence Principle. The acceleration could also be used to cancel an existing gravitational field—for example, inside a freely falling elevator passengers are weightless, conditions are equivalent to those in the unaccelerated space station in outer space.

It is important to realise that this equivalence between a gravitational field and acceleration is only possible because the gravitational mass is exactly equal to the inertial mass. There is no way to cancel out electric fields, for example, by going to an accelerated frame, since many different charge to mass ratios are possible.

As physics has developed, the concept of fields has been very valuable in understanding how bodies interact with each other. We visualize the electric field lines coming out from a charge, and know that something is there in the space around the charge which exerts a force on another charge coming into the neighborhood.

We can even compute the energy density stored in the electric field, locally proportional to the square of the electric field intensity. It is tempting to think that the gravitational field is quite similar—after all, it's another inverse square field. Evidently, though, this is not the case. If by going to an accelerated frame the gravitational field can be made to vanish, at least locally, it cannot be that it stores energy in a simply defined local way like the electric field.

We should emphasize that going to an accelerating frame can only cancel a *constant* gravitational field, of course, so there is no accelerating frame in which the whole gravitational field of, say, a massive body is zero, since the field necessarily points in different directions in different regions of the space surrounding the body.

SOME CONSEQUENCES OF THE EQUIVALENCE PRINCIPLE

Consider a freely falling elevator near the surface of the earth, and suppose a laser fixed in one wall of the elevator sends a pulse of light horizontally across to the corresponding point on the opposite wall of the elevator. Inside the elevator, where there are no fields present, the environment is that of an inertial frame, and the light will certainly be observed to proceed directly across the elevator. Imagine now that the elevator has windows, and an outsider at rest relative to the earth observes the light.

As the light crosses the elevator, the elevator is of course accelerating downwards at g, so since the flash of light will hit the opposite elevator wall at precisely the height relative to the elevator at which it began, the outside observer will conclude that the flash of light also accelerates downwards at g.

In fact, the light could have been emitted at the instant the elevator was released from rest, so we must conclude that light falls in an initially parabolic path in a constant gravitational field. Of course, the light is traveling very fast, so the curvature of the path is small! Nevertheless, the Equivalence Principle forces us to the conclusion that the path of a light beam is bent by a gravitational field.

The curvature of the path of light in a gravitational field was first detected in 1919, by observing stars very near to the sun during a solar eclipse.

The deflection for stars observed very close to the sun was 1.7 seconds of arc, which meant measuring image positions on a photograph to an accuracy of hundredths of a millimeter, quite an achievement at the time. One might conclude from the brief discussion above that a light beam in a gravitational field follows the same path a Newtonian particle would if it moved at the speed of light. This is true in the limit of small deviations from a straight line in a constant field, but is not true even for small deviations for a spatially varying field, such as the field near the sun the starlight travels through in the eclipse experiment mentioned above.

We could try to construct the path by having the light pass through a series of freely falling elevators, all falling towards the centre of the sun, but then the elevators are accelerating relative to each other (since they are all falling along *radii*), and matching up the path of the light beam through the series is tricky.

If it is done correctly it turns out that the angle the light beam is bent through is twice that predicted by a naove Newtonian theory.

What happens if we shine the pulse of light vertically *down* inside a freely falling elevator, from a laser in the centre of the ceiling to a point in the centre of the floor? Let us suppose the flash of light leaves the ceiling at the instant the elevator is released into free fall. If the elevator has height h, it takes time h/c to reach the floor. This means the floor is moving downwards at speed gh/c when the light hits.

Question: Will an observer on the floor of the elevator see the light as Doppler shifted? The answer has to be no, because inside the elevator, by the Equivalence Principle, conditions are identical to those in an inertial frame with no fields present. There is nothing to change the frequency of the light. This implies, however, that to an outside observer, stationary in the earth's gravitational field, the frequency of the light *will* change.

This is because he will agree with the elevator observer on what was the initial frequency f of the light as it left the laser in the ceiling (the elevator was at rest relative to the earth at that moment) so if the elevator operator maintains the light had the same frequency f as it hit the elevator floor, which is moving at gh/c relative to the earth at that instant, the earth observer will say the light has frequency $f(1 + v/c) = f(1+gh/c^2)$, using the Doppler formula for very low speeds.

We conclude from this that light shining downwards in a gravitational field is shifted to a higher frequency. Putting the laser in the elevator floor, it is clear that light shining upwards in a gravitational field is red-shifted to lower frequency. Einstein suggested that this prediction could be checked by looking at characteristic spectral lines of atoms near the surfaces of very dense stars, which should be red-shifted compared with the same atoms observed on earth, and this was confirmed.

This has since been observed much more accurately. An amusing consequence, since the atomic oscillations which emit the radiation are after all just accurate clocks, is that time passes at different rates at different altitudes. The US atomic standard clock, kept at 5400 feet in Boulder, gains 5 microseconds per year over an identical clock almost at sea level in the Royal Observatory at Greenwich, England. Both clocks are accurate to one microsecond per year. This means you would age more slowly if you lived on the surface of a planet with a large gravitational field. Of course, it might not be very comfortable.

PARTICLE CREATION

A synchrocyclotron, it is possible to accelerate protons to relativistic speeds. The rest energy of a proton $m_p c^2$ is 938 MeV, using here the standard high energy physics energy unit: 1 MeV = 10^6 eV. The neutron is a bit heavier—$m_n c^2$ = 940 MeV. (The electron is 0.51 MeV). Thus to accelerate a proton to relativistic speeds implies giving it a K.E. of order 1, 000 MeV, or 1 GeV.

The standard operating procedure of high energy physicists is to accelerate particles to relativistic speeds, then smash them into other particles to see what happens. For example, fast protons will be aimed at protons at rest (hydrogen atoms, in other words—the electron can be neglected). These proton-proton collisions take place inside some kind of detection apparatus, so the results can be observed.

One widely-used detector is the bubble chamber: a transparent container filled with a superheated liquid. The electric field of a rapidly moving charged particle passing close to a molecule can dislodge an electron, so an energetic particle moving through the liquid leaves a trail of ionized molecules. These give centers about which bubbles can nucleate. The bubbles grow rapidly and provide a visible record of the particle's path. What is actually observed in *p-p* scattering at relativistic energies is that often more particles come out than went in—particles called pions, π^+, π^0, π^- can be created. The π^0 is electrically neutral, the π^+ has exactly the same amount of charge as the proton. It is found experimentally that total electric charge is always conserved in collisions, no matter how many new particles are spawned, and total baryon number (protons + neutrons) is conserved.

Possible scenarios include:

$$p + p \rightarrow p + p + \pi^0$$

and

$$p + p \rightarrow p + n + \pi^+.$$

The neutral pion mass is 135 MeV, the charged pions have mass 140 MeV, where we follow standard high energy practice in calling mc^2 the "mass", since this is the energy equivalent, and hence the energy which, on creation of the particle in a collision, is taken from kinetic energy and stored in mass. However, an incoming proton with 135 MeV of kinetic energy will not be able to create a neutral pion in a collision with a stationary proton. This is because the incoming proton also has momentum, and the collision conserves momentum, so some of the particles after the collision must have momentum and hence kinetic energy.

The simplest way to figure out just how much energy the incoming proton needs to create a neutral pion is to go to the centre of mass frame, where initially two protons are moving towards each other with equal and opposite velocities, there is no total momentum. Obviously, in this frame the least possible K.E. must be just enough to create the π^0 with all the final state particles (p, p, π^0) at rest. Thus if the relativistic mass of the incoming protons in the centre of mass frame is m, the total energy

$$E2mc^2 = 2m_pc^2 + m_zc^2.$$

Putting in the proton and pion masses from above, and using $m = m_p/\sqrt{1-(v^2/c^2)}$, we find the two incoming protons must both be traveling at $0.36c$.

Recall that this is the speed in the centre of mass frame, and for practical purposes, like designing the accelerator, we need to know the energy necessary in the "lab" frame—that in which one of the protons is initially at rest. The two frames obviously have a relative speed of $0.36c$, so to get the speed of the incoming proton in the lab frame we must add a velocity of $0.36c$ to one of $0.36c$ using the relativistic addition of velocities formula, which gives $0.64c$. This implies the incoming proton has a relativistic mass of 1.3 times its rest mass, and thus a K.E. around 280 MeV.

Thus to create a pion of rest energy 135 MeV, it is necessary to give the incoming proton at least 290 MeV of kinetic energy. This is called the "threshold energy" for pion production. The "inefficiency" arises because momentum is also conserved, so there is still considerable K.E. in the final particles.

Antiproton Production

On raising the energy of the incoming proton further, more particles are produced, including the "antiproton"—a negatively charged heavy particle which will annihilate a proton in a flash of energy. It turns out experimentally that an antiproton can only be produced accompanied by a newly created proton,

$$p + p \rightarrow p + p + p + \overline{p}$$

Notice we could have conserved electric charge with less energy with the reaction

$$p + p \rightarrow p + p + \pi^+ + \overline{p}$$

but this doesn't happen—so charge conserv+ation isn't the only constraint on which particles can be produced. In fact, what we are seeing here is experimental confirmation that the conservation of baryon number, which at the low energies in the context of pion production just meant that the total number of protons plus neutrons stayed fixed, is generalized at high energies to include antiparticles with negative baryon number, -1 for the antiproton. Thus baryon number conservation becomes parallel to electric charge conservation, new particles can always be produced at high enough energies provided the total new charge and the total new baryon number are both zero. We should emphasize again that these are *experimental* results gathered from examining millions of collisions between relativistic particles.

One of the first modern accelerators, built at Berkeley in the fifties, was designed specifically to produce the antiproton, so it was very important to calculate the antiproton production threshold correctly! This can be done by the same method we used above for pion production, but we use a different trick here which is often useful. We have shown that on transforming the energy and momentum of a particle from one frame to another

$$E^2 - c^2p^2 = E'^2 - c^2p'^2$$

Since the Lorentz equations are linear, if we have a system of particles with total energy E and total momentum p in one frame, E', p' in another, it must again be true that

$$E^2 - c^2\vec{p}^2 = E'^2 - c^2\vec{p}'^2.$$

We can use this invariance to get lab frame information from the centre of mass frame. Noting that in the centre of mass (CM) frame the momentum is zero, and in the lab frame the momentum is all in the incoming proton,

$$E_{cm}^2 = ((m_{in} + m_0)c^2)^2 - c^2p_{in}^2$$

where here m_0 is the proton rest mass, m_{in} is the relativistic mass of the incoming proton. At the antiproton production threshold, $E_{cm} = 4m_0c^2$, so

$$16_0^2c^4 = m_{in}^2c^4 + 2m_{in}c^2m_0c + m_0^2c^4 - c^4p_{in}^2$$

and using

$$m_{in}^2c^4 - c^2p_{in}^2 = m_0^2c^4$$

we find

$$2(m_{in}c^2)(m_0c^2) + 2(m_0c^2)^2 = 16(m_0c^2)^2,$$

so

$$m_{in}c^2 = 7m_0c^2.$$

Therefore to create two extra particles, with total rest energy $2m_0c^2$, it is necessary for the incoming proton to have a kinetic energy of $6m_0c^2$. The Berkeley Gevatron had design energy 6.2 GeV. As we go to higher energies, this "inefficiency" gets worse—consider energies such that the kinetic energy >> rest energy, and assume the incoming particle and the target particle have the same rest mass, m_0, with the incoming particle having relativistic mass m.

Comparing the centre of mass energy with the lab energy at these high energies,

$$E_{LAB} = (m + m_0)c^2,$$

$$E_{CM}^2 = E_{LAB}^2 - p^2_{LAB}c^2$$

$$= m^2c^4 + 2mc^2 + m_0^2c^4 - p^2_{LAB}c^2$$

$$= 2m_0c^2 + (mc^2 + m_0c^2).$$

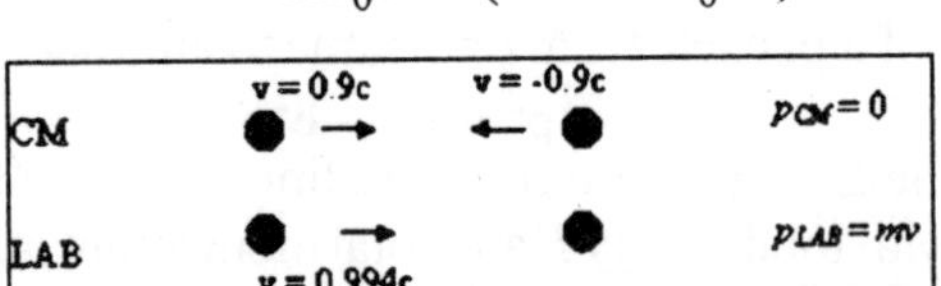

Fig. The Same Collision as Viewed in the CM and LAB Frames

For $m >> m_0$,

$$E_{CM}^2 \approx 2m_0c^2mc^2 \approx 2m_0c^2.E_{LAB}$$

so

$$E_{CM} \approx \sqrt{2m_0c^2.E_{LAB}},$$

ultimately one must *quadruple* the lab energy to double the centre of mass energy.

GENERAL RELATIVITY AND THE GLOBAL POSITIONING SYSTEM

Despite what you might suspect, the fact that time passes at different rates at different altitudes has significant practical consequences. An important *everyday* application of general relativity is the Global Positioning System. A GPS unit finds out where it is by detecting signals sent from orbiting satellites at precisely timed intervals. If all the satellites emit signals simultaneously, and the GPS unit detects signals from four different satellites, there will be three relative time delays between the signals it detects.

The signals themselves are encoded to give the GPS unit the precise position of the satellite they came from at the time of transmission. With this information, the GPS unit can use the speed of light to translate the detected time delays into distances, and therefore compute its own position on earth by triangulation.

But this has to be done very precisely. Bearing in mind that the speed of light is about one foot per nanosecond, an error of 100 nanoseconds or so could, for example, put an airplane off the runway in a blind landing. This means the clocks in the satellites timing when the signals are sent out must certainly be accurate to 100 nanoseconds a day. That is one part in 10^{12}. It is easy to check that both the special relativistic time dilation correction from the speed of the satellite, and the general relativistic gravitational potential correction are much greater than that, so the clocks in the satellites must be corrected appropriately.

The satellites go around the earth once every twelve hours, which puts them at a distance of about four earth radii. The calculations of time dilation from the speed of the satellite, and the clock rate change from the gravitational potential, are left as exercises for the student. In fact, Ashby reports that when the first Cesium clock was put in orbit in 1977, those involved were sufficiently skeptical of general relativity that the clock was not corrected for the gravitational redshift effect.

But just in case Einstein turned out to be right the satellite was equipped with a synthesizer that could be switched on if necessary to add the appropriate relativistic corrections. After letting the clock run for three weeks with the synthesizer turned off, it was found to differ from an identical clock at ground level by precisely the amount predicted by special plus general relativity, limited only by the accuracy of the clock.

This simple experiment verified the predicted gravitational redshift to about one per cent accuracy. The synthesizer was turned on and left on.

Index